TK 4035 .M37 C58 1998

Clark, William H.,

Electrical design guide for
commercial buildings

DATE DUE

DEMCO 38-297

D1566247

NEW ENGLAND INSTITUTE
OF TECHNOLOGY
LEARNING RESOURCES CENTER

Electrical Design Guide for Commercial Buildings

Electrical Design Guide for Commercial Buildings

William H. Clark II, P.E.

NEW ENGLAND INSTITUTE
OF TECHNOLOGY
LEARNING RESOURCES CENTER

McGraw-Hill

New York   San Francisco   Washington, D.C.   Auckland   Bogotá
Caracas   Lisbon   London   Madrid   Mexico City   Milan
Montreal   New Delhi   San Juan   Singapore
Sydney   Tokyo   Toronto

McGraw-Hill
A Division of The **McGraw-Hill** Companies

Copyright © 1998 by The McGraw-Hill Companies, Inc. All rights reserved. Printed in the United States of America. Except as permitted under the United States Copyright Act of 1976, no part of this publication may be reproduced or distributed in any form or by any means, or stored in a data base or retrieval system, without the prior written permission of the publisher.

1 2 3 4 5 6 7 8 9 0   DOC/DOC   9 0 3 2 1 0 9 8

ISBN 0-07-011991-0

National Electrical Code® and NEC® are registered trademarks of the National Fire Protection Association, Inc., Quincy, MA 02269.

The sponsoring editor for this book was Zoe G. Foundotos, the editing supervisor was Scott Amerman, and the production supervisor was Pamela Pelton. It was set in Times Roman by Ron Painter of McGraw-Hill's Professional Book Group composition unit.

Printed and bound by R. R. Donnelley & Sons Inc.

McGraw-Hill books are available at special quantity discounts to use as premiums and sales promotions, or for use in corporate training programs. For more information, please write to the Director of Special Sales, McGraw-Hill, 11 West 19th Street, New York, NY 10011. Or contact your local bookstore.

Information contained in this work has been obtained by The McGraw-Hill Companies, Inc. ("McGraw-Hill") from sources believed to be reliable. However, neither McGraw-Hill nor its authors guarantees the accuracy or completeness of any information published herein and neither McGraw-Hill nor its authors shall be responsible for any errors, omissions, or damages arising out of use of this information. This work is published with the understanding that McGraw-Hill and its authors are supplying information but are not attempting to render engineering or other professional services. If such services are required, the assistance of an appropriate professional should be sought.

 This book is printed on recycled, acid-free paper containing a minimum of 50% recycled, de-inked fiber.

Contents

Foreword   xi

Chapter 1. The Fundamentals of Electricity   1

    Engineering Analogies   2
    Conductors   2
    Voltage Drop   7
    Resistance Losses   8
    Circuit Elements   9
    Insulators   11
    Plastic Insulated Sheathed Cables   11
    Feeder Size Calculations   12

Chapter 2. Electrical Circuits   19

    Ohm's Law   19
    Current in a Series Circuit   21
    Parallel Circuits   22
    Series-Parallel Circuits   24
    Whetstone Bridges   25
    Adjusting Bridge Sensitivity   26
    Kirchoff's Laws   27
    Kirchoff's Current Law   28
    Kirchoff's Voltage Law   28
    Three-Phase Power and Neutral Current   29
    Alternating Current   30

Chapter 3. Phase Diagrams   31

    Sinusoidal Waveforms   31
    Inductors   33
    Capacitors   33
    Impedance in Circuits   35
    Reactance on the Imaginary Plane   36
    Real and Reactive Power   37
    The Frequency Domain   39

vi   Contents

    Power Factor Correction   41
    Safety   41

Chapter 4. Power Circuiting   43

    Temporary Service   43
    Service Installation   43
    Points of Use   52
    One-Line Riser Diagrams   55
    Schedules   57
    Phasor Representations   61

Chapter 5. Signals Circuiting   65

    Exit Signs   65
    Telephone and Cable   66
    Communication Rooms   68
    Fire Alarm Devices   69
    ADA Devices   71
    Computer Networking   72
    Fiber Optic Cabling   72
    Security Devices   74

Chapter 6. Lighting and Circuiting   77

    Luminaire and Lamp Selections   78
    The Reflected Ceiling Plan   80
    Special Areas   83
    Circuiting and Switching   84
    Conserving Energy   87
    Electromagnetic Theory   88

Chapter 7. Power Distribution System   91

    Operations Research   91
    The Design Team   92
    Decision Making   93
    The Big Picture for Power   95
    Three-Phase Zoning   95
    Lighting   98
    Demand Factors   99
    Conduit Supports   100
    The Service Load   102

Chapter 8. Designing the Visual Environment   103

    Maximizing Impact   103
    Lamp Selection   105

Designing the Visual Environment	105
Identify a Fixture Budget	106
The Lighting Palette	107
Daylighting Strategies	108
Light Shelf	109
The Nature of Light	110
Zonal Cavity Calculations	111
Levels of Illumination	113

Chapter 9. HVAC Equipment — 119

Packaged Rooftop Equipment	124
Split Systems	125
Air Handler Rooms	126
Variable-Frequency Drives	127
Central Plant	128
Cooling Tower	129
Miscellaneous Equipment	129
Integrated Systems	135
Individual Control Elements	138
Grounding Conductors	139
Overload Protection	142

Chapter 10. Motors — 143

Generator Sizing	143
Motor Coordination	151
Motors in a Mechanical System	152
Operating Parameters	154
Pump Curves	162
Motor Circuiting	167

Chapter 11. Transformers — 169

Core Losses	170
Applications	173
Three-Phase Connections and Polarity	174
Isolation Transformers	178

Chapter 12. Estimating — 179

Service Corridors	180
Lighting Fixtures	181
Devices	181
Motors	182
Conduit Costs and Labor	189
Additional Factors	189

Chapter 13. Computer Applications — 195

- The Platform — 196
- The Data Entry Form — 198
- The Program — 200

Chapter 14. Value Engineering — 209

- The Process — 211
- The State versus the Architectural/Engineering (A/E) Firm — 211
- Costs versus Functions — 213
- Diagram of the Plan — 214
- Individual Deliberations — 216
- Brainstorming — 216
- Evaluation — 217
- Project Evaluations — 220
- The Presentation — 224
- Lessons for In-House VE Programs — 226
- Value Engineering Smaller Projects — 228

Chapter 15. Load Control and Harmonics — 235

- Electrical Load Profile Elements — 236
- Proactive Controls — 237
- Lighting — 238
- Motors — 239
- Equipment — 240
- Electric Heating — 240
- Power Quality — 241
- Symptoms — 243
- Power Conditioning — 243
- Transformers — 245
- Power Producers — 246
- Synchronous Motors — 248

Chapter 16. Special Applications — 251

- Water Source Heat Pumps — 252
- Electrical Advantages — 252
- Outside Air Systems — 255
- Hybrid Systems — 256
- Complex Systems — 258
- Balanced Electricity Use — 260
- Electric Power at Sea: Surface Combatants — 260
- Collateral Issues — 261
- HVAC Systems — 262
- The Equipment — 262
- Application of the Technology — 263
- Electrical Systems at Sea — 264

Distributed Power Control	265
Harmonic Power Distortions	266
Communications Backbone	266
Central Energy Management System	266
Oil Field Applications	266
Hot Water Usage	267
Peak Demand Reduction	268

Chapter 17. Designing in CAD Programs — 269

External Referencing	270
Layer Names	271
Line Types	273
Detail Drawings	274
General Notes	277

Chapter 18. Short Specifications — 281

Chapter 19. Electrical Checklist — 299

Division 16—Electrical Specifications	302
Division 0 Sections/Checklist	308
General Requirements	309

Chapter 20. Designing for Conservation — 311

Lighting Power Limits	311
The Method	312
Rebate Programs	312

Chapter 21. The Production Process — 323

The Schematic Phase	324
Design Development	328
Bidding	332
Addenda	334
Value Engineering	334
Construction Management	334
Change Orders	335

Glossary   339

Index   351

Foreword

This book is structured to provide insights and guidance to individuals with some prior experience in electrical design or contracting. The very basic concepts and principles are not included, not because they aren't quite important, but to allow room for some more pertinent and valuable concepts and design practices. In addition the text strives to endow the reader with a better "feel" or understanding of some of the important facets of the trade. An individual can very easily complete a whole career in the trade with only a rudimentary understanding of such basic ideas as resistance, capacitance, and inductance, not to mention three-phase alternating current and harmonics. However, a little better comprehension of these very important (and actually quite complex) concepts will make for a much more interesting career. And a safer one, too.

Another important goal is to suggest additional tools and techniques so that better and less costly electrical installations can be designed and installed. The ultimate objective is the conservation of material and human resources as well as electrical energy used after completion, all of which inevitably arise from quality design. Unlike other disciplines, in which energy conservation is a simple matter of purchasing the most efficient equipment, electrical designers must achieve energy conservation through whole systems rather than parts of the system. Of course, some electrical items are more efficient than others (notably in lighting), yet the greatest savings are accomplished via rigorous and careful design practices as they apply to the entire power grid.

Electrical systems are more than just a costly, complex part of any building; they are a distribution network that delivers power to all parts of a building. This distribution is not done without losses, however. Power is lost in roughly equal proportions to resistance in wire and connectors, nonunity power factors, and harmonics. Minimizing these losses is a difficult design problem, and it is only done through exacting attention to the design of every single element of the whole system. Diligence in the designing, then, will not only reduce energy consumption at the facility but also substantially reduce construction costs and result in a much safer, more adaptable electrical system.

The Audience

As noted earlier, this is not an introductory book. It assumes the reader already has a basic understanding of electrical devices. The ambitious electrical journeyman, for example, should find much to enhance his or her understanding of the engineering that goes into the systems being installed. So will the master electrician, who may be challenged by increasingly complex projects such as electrical systems at hospitals and computerized industrial facilities. These are the most profitable jobs in the industry, and if they cannot be done with confidence, then the growth potential of an individual or a contracting firm is greatly diminished. Finally, for the general contractor, there are new standards and procedures that can help to control the bidding, construction, and troubleshooting aspects of the business. In all cases the result of a little conscientious study is greater job satisfaction and safety, a higher quality of workmanship, and more satisfied clients. All of which ultimately mean enhanced productivity, plus repeat business.

Professionals should read the book too. There are mechanical engineers licensed in hvac needing to seal the complete set of engineering drawings on small jobs in order to meet their profit margins. And the nonelectrical engineering students, too—all of whom are required to take an introductory course in "EE"—they should be familiar with the contents of this book. They can learn much more practical knowledge in a design-oriented text such as this than from the typical theoretical tome. Such a heavy coverage has no information past the first few chapters that the engineer will ever use again, and the remainder is so abstract and impractical that it leaves the student with a lifelong apprehension of anything having to do with electricity.

Then there are newly graduated electrical engineers, working for consulting firms and planning to become registered. They will find important guidance in design issues that are not often taught in college, ones that permeate the practice of engineering once you are qualified to seal drawings but which are rarely treated thoroughly by senior engineers. These issues may very well be so new (e.g., harmonics and computer power supply requirements) that even the very experienced engineers are mostly ignorant of them. This realm, once the exclusive concern of hospital and other specialty equipment designers, is now a part of almost every commercial job because of the widespread use of computers and microelectronic devices.

Experienced electrical engineers will find much to learn, too. The book is up to date on the issues of harmonic distortion in power supplies, methods of computer networking design, computer-aided drafting, and software design capabilities. These and other important developments in just the last few years are of vital concern to the engineer's success—and stress level—on almost any project now. Better to spend a little time learning from a book what you need to know now, instead of learning it when change orders come in from the field or a summons comes from the owner's attorney.

Not that engineers are the only individuals who need to read this book. Architects stand to find help, for one thing, in circuiting the very small jobs that are all too often not designed at all. Creating a more detailed bid document set reduces cost and minimizes conflicts during the construction phase. Even more importantly, on the larger jobs the architect in the know can spot instances when the project engineer is asking for more than is necessary—space, money, or design time—and so, keep project scheduling and costs under strict control.

Lighting designers must be included as readers. They can learn enough to lay out the entire power distribution system for their designs. This skill will make for a more thorough and comprehensive power distribution system and reduce production time and costs. Lighting consultants will be able to create designs that are not only less expensive but which provide a more uniform level of lighting. Reducing cost is especially important, as any money saved on either the production or installation will increase the funds available for the lighting system itself. This is always a plus in the aesthetic design of any space or structure.

Even construction superintendents will benefit. Electricians are notoriously circumspect about their means and methods—all the while demanding the highest wages of the trades. There is no better way to keep them in line, on schedule and in budget, than to bolster your understanding of what they do and why they do it. Engineering designers are not much better—if you know for a fact it takes only ten minutes to make a design decision, you can keep from being delayed for a week or billed for a day's work. The more you know, the less inclined people will be to overcharge or mislead you.

Finally, there are the energy auditors. Traditionally the only role electrical systems play in reducing energy consumption is by minimizing the consumption by lighting. This is no longer the case. As semiconductor-based systems continue to proliferate, harmonics escalate, neutral current increases, and, thus, needless power losses grow exponentially. Each of these problems is treated thoroughly, with the goal of designing for energy efficiency. Other typical design techniques, such as designing for a 2 percent voltage drop, are analyzed in the light of energy use, with solid recommendations on conscientious principles for designing systems of superior quality and longevity.

The Perspective

All of the comments thus far have been directed at professionals practicing in well-regulated construction environments. In the United States, for example, there are many levels of oversight. First, there is the National Electrical Code (NEC) which is developed by the industry as a standard for all to practice. Federal installations do not necessarily have to comply with the NEC, although it is a rare instance when its guidelines are knowingly violated. All other state and local jurisdictions adopt the NEC as their code, with some local provisions often promulgated in deference to the local political environment. For example, in the interests of reducing energy use, certain power limits may be imposed on lighting, or efficiency limits may be established for motors and hvac equipment. Or specific measures to enhance the safety of the occupants (and the protection of the utility's equipment) may be required, such as a main service disconnect. This sort of requirement is most common in areas with a city-owned utility, whereby electrical design rules can be passed and enforced since they have responsibility for the electrical service and its regulation.

In addition to or sometimes instead of these rules, specific organizations may have developed their own design standards. Federal agencies such as the Department of Veterans Affairs and the Defense Department have completely different standards than almost anyone else. At one time these standards might have been higher than the industry as a whole; thus they would have served an important purpose. Now they are often below the high standards of the NEC, and as a result they not only permit unsafe conditions but cost the agencies more. The reason for this situation occurring is that it is hard enough already to keep up with the changing code criteria. Add a completely new code, and consultants must by necessity charge a premium to keep up to date with that code too. Few firms are willing to do this. This reduces the competition for the design of projects, thereby further increasing the cost to the government. Some state agencies have the same immunity—and its concommitant cost escalation.

Much the same happens with the contracting side of the situation, with all the trades in addition to electrical. There is a new standard of work with a lot more oversight by the owner (i.e., the government entity), which slows down the entire process and generates mountains of paperwork. The more documentation that is generated, the more time consuming and costly the project is, and the greater the likelihood that general contractors would rather spend their energy and resources on commercial projects that can be run more expeditiously and efficiently.

There is an important lesson in this situation, for less-developed nations or regions with ambitions of growing into a stronger prosperity.

Free enterprise has a wonderful way of creating products as quickly and inexpensively as practicable. Codes that bear upon the engineering disciplines (e.g., the NEC) are needed to ensure a safe installation. Further constraints beyond issues of health and safety, and the bogie of diminishing returns, drives costs up quite quickly. There is really no other way.

Finally, it is important to understand that local conditions can limit the usefulness of an engineering design standard. If the products, the craftspeople, and the funds are not available for proper implementation of an all-encompassing code, then having such a code may do more harm than good. The more rules that cannot be uniformly and fairly enforced, the more influence that gives the regulating authorities to enforce the rules unfairly, unevenly, and sometimes even with the prerequisite of graft. It is better, for developing areas, to have a fundamental code that challenges all to create and build to a higher, but achievable, standard. Once these basic standards become common practice, then the standards can be raised, and so forth until eventually a consistently high quality of work is everywhere in evidence.

This book is an example of a basic code, covering all aspects of the industry, that can be safely and confidently promulgated in lesser-developed areas.

The Competition

The design-build concept has taken the consulting engineering field by surprise. By some estimates over 60 percent of all projects are now done by the design-build process, in which contractors and engineers create a team to bid for the design and construction of a project. Generally, this means the actual design documents have to be less complete than for bid projects, since much of the design is done in the field.

Not that a great deal of the actual design is not already done in the field, even on conventional projects for which a complete engineering package is generated for bidding and construction by others. This is especially true of the electrical portion of the project, for which many of the drawings are only schematics anyway. In addition, those portions that are designed—such as power or lighting circuiting and panel schedules—are inevitably redesigned by the field personnel, which results in the most efficient use of material resources such as conduit, wiring, and junction boxes.

This book endeavors to bring the engineer's design closer to what is actually installed, so that the electrical contractor must do less of a redesign job. At the same time, it provides the contractor with the skills to do the engineer's job to a greater and better extent. Some sections are addressed to the engineer, others to the designer or contractor, depending on what aspect of the project is customarily done by the specific trade. This approach may at first seem a little confusing, unless you are that individual who wears several hats and does the entire project.

Such a role is that of the design-build contractor; this book is written for that individual. It takes the middle manager out of the loop, downsizing the entire process by transferring all of the design engineer's duties to the contractor. This is the basis for the design-build concept. It makes the process more efficient by streamlining, and the traditional role of the consulting engineer is eliminated entirely.

On small commercial and residential projects, the architect already does the majority of the electrical design necessary for permitting and bidding. On light commercial projects the architect teams with an electrical contractor to do so. It is only on the very large projects that a consulting firm is involved any more, and it is only a matter of time before the larger contracting firms will have all the skills to do the job more efficiently, cheaply, and effectively. This book is intended to help them toward that end.

Conversely, it is a warning to consultants that unless they upgrade their skills to create a final product that is more useful to contractors—less schematic and more like the final product—they may lose their job entirely to the design builders. The same has happened to others in the trade, notably hvac designers who draw two-line ducts for everything instead of the more schematic one-line. Plumbers must not only draw all lines in their plans but also provide very detailed riser diagrams. Only by providing such valuable equivalents in their own drawings will electrical engineers stay in business.

This additional detail work, of course, will take a great deal of extra time, and the fee will probably remain the same. So a concerted effort is required to make the whole electrical engineering design process more efficient and effective. This book addresses a number of issues such as upgrades or modifications to traditional design processes, many of which are rooted in the old manual drafting means and methods. The advantages of modern, effective computer-aided drafting and computer design programs are that they are integrated into an overall design approach to result in a more complete, accurate, and reality-based electrical design. Both of these electronic means can save huge blocks of time in the design and drafting process. This extra time can be put to good use: designers can create a more comprehensive system design, one in keeping with the highest standards of sturdy, flexible, cost-effective, and energy-efficient designs that have always been expected of engineering consultants.

Electrical Design Guide for Commercial Buildings

Chapter

1

The Fundamentals of Electricity

A little time spent to become thoroughly familiar with the basic principles of electricity is always time well spent. Even the most seasoned professionals need to refresh their understanding now and then to become current on the trade.

The typical designer, engineer, and contractor spends most of his or her time remote from anything akin to true electrical design. Mostly the work is involved with strict code interpretation. It is assumed that all of the important rules for the behavior of electrical current are built in to the codes, so it is not necessary to go any further than to ensure compliance. This is the standard set by most regulatory authorities, inspectors, and legal advocates. If a design can be shown to comply with the code enforced at the time, the designer or engineer is legally without any complicity in the event problems occur.

The codes do in fact follow the principles of electricity, but it is wrong to assume that the codes parallel closely the laws of physics or that compliance with the industry standards will keep any design from having problems. In fact the codes are based upon a very conservative application of the principles of electricity. Often the safety factor is so high, presumably to keep the compliance authorities themselves out of trouble, that the cost to the user is prohibitive. This can result in such strict fiscal constraints that substandard materials and devices must be used or underdesigned systems installed. In either case a well-informed electrical designer can size up the situation accurately, make a responsible judgment call on the installation, and arrive at a proper balance between the guidance of general, written codes and the budget and the specific operating characteristics of the system in operation. Such a design is based to a large extent upon the designer's confidence in the fundamentals of electricity. It will be a safer and more trouble-free design than might have been devised by someone mindful only of strict compliance to the codes, who does no more than just follow the permitting process of the local regulating authority.

Engineering Analogies

Electricity is an abstract phenomenon, not easily rationalized because its effect is so far from everyday perceptions. Generally, people know what it does, that some large appliances use a lot of it, and that it can cause great harm. Otherwise it is best left to the experts, even though the experts may have only a little better feel for it.

There are simple analogies in the physical world that can help strengthen the understanding of electricity and its behavior. These are actually a lot more than analogies. The mathematical expression of the phenomena (i.e., the laws of physics that are used to predict and quantify behavior) are exactly the same. Thus, the correlation is exact. The comparison of electricity flow in wiring is equivalent to the flow of water through piping.

Electrical constant	Fluid constant
Movement of electrons	Flow of water
Metal wire in a sheath	Volume defined by the pipe walls
Amps of current	Gallons per minute of flow
Resistance losses	Resistance due to rough pipe ID or mineral deposits
Voltage	Head of pressure needed to move the fluid

From these principles, there is the understanding that the larger the pipe the more water can be moved with less static pressure losses due to pipe friction; the larger the wire the more current can pass with a lower resistance loss. Also, for a given voltage, or the same size pump, more amps may flow in the larger wire with a greater voltage difference at the end of the run, and more pressure is available in the hydronic system with larger pipes.

A little more sophisticated comparison looks at control elements in the systems. A capacitor in an electrical system stores charge and modulates the extremes, much like a spring in the suspension of a vehicle or a shock absorber in a plumbing system to reduce water hammer.

Conductors

At the macroscopic level, then, electric current flows through wiring in very much the same sense as water flows through piping. This flow is subject to physical laws, and these constraints are what allow its behavior to be monitored, controlled, and predicted. In fact, these defining limits predicate the constraints in some cases. The flow of electricity is, for example, restricted to a medium such as a conductor. The properties of this conductor, in turn, determine the amount of current that can flow and the resistance to flow that the material exhibits.

Table 1.1 lists the properties of some common conductors. Figure 1.1

TABLE 1.1 Properties of Conducting Metals

Material	Description	Temperature coefficient	Specific resistance, (cmil-Ω/ft)
Aluminum	Element (Al)	0.004	17
Carbon	Element (Cu)	−0.0003	*
Constantan	55% Cu 45% Ni	0	295
Copper	Element (Cu)	0.004	10.4
Gold	Element (Au)	0.004	14
Iron	Element (Fe)	0.006	58
Manganin	84% Cu, 12% Mn, 4% Ni	0	270
Nichrome	65% Ni 23% Fe 12% Cr	0.0002	676
Nickel	Element (Ni)	0.005	52
Silver	Element (Ag)	0.004	9.8
Steel	99.5% Fe, 0.5% C	0.003	100
Tungsten	Element (W)	0.005	33.8

*The resistance of carbon is 2500 to 7500 times the resistance of copper.

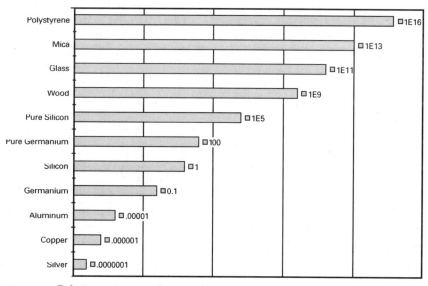

Figure 1.1 Relative resistance of materials.

gives a graphical presentation of some common materials. The values are for specific resistance, measured in a 1-ft length of wire with a cross sectional area of 1 circular mil (cmil). The lower the resistance, the better the conductor. Silver has the lowest resistance, followed by copper, gold, and aluminum. Other materials listed have much higher values—by several orders of magnitude—but are still metals.

As implied by the fluid analogy, the resistance to flow is also proportional to the size of the path, or wire in the case of current carrying metal. The formula for this relationship is

$$R = \frac{\rho}{A} \qquad (1.1)$$

where R = total resistance
ρ (rho) = specific resistance, or resistivity
A = cross-sectional area of the wire

The area of round wire is measured in circular mils. A mil is one-thousandth of an inch, 0.001 in, and 1 cmil is the cross section of a wire with a diameter of 1 mil. In different terms, this means that the square of the diameter of any circular area in mils is equal to the number of circular mils in the area. This relationship between diameter in mils and area in circular mils is illustrated in Table 1.2. The resistance in ohms of copper wire is also provided to show how resistance varies with length of the conductor. The first column of the table gives the gage number, a standardized round wire size called American Wire Gage (AWG). Note that the area of the wire doubles for every three gage sizes and that as the wire size increases the resistance decreases.

Another important characteristic of conductors is their sensitivity to temperature. When a feeder or home run is fully loaded, or overloaded if there are harmonics present, the resistance to current flow is evidenced as thermal energy: The wire heats up. Many common circumstances can cause the wire to be loaded beyond its design capacity, such as a motor running beyond its rating, extra appliances wired to the circuit, or wire routed in a 150°F attic, across a hot, black roof, or near a boiler flue or other heat source. An appreciation for the effects of temperature should encourage the electrician to design against them.

Mathematically, the relationship between heat and resistance is an exponential function. As the temperature of the conductor increases, the resistance increases at an exponential rate. At temperatures near ambient, the function can be accurately represented by the linear function

$$R_t = R_o \times (\alpha \Delta t) \qquad (1.2)$$

where R_o = resistance at 20°C
R_t = resistance at the higher temperature
ΔT = temperature rise above ambient
α = temperature coefficient, from Table 1.1

Consider a copper wire in the attic of a home, where the temperature can be 50°C above ambient. If the wire is #12 AWG, and the initial resistance from Table 1.1 is 1.619, the resistance at the elevated temperature is

TABLE 1.2 Copper Wire Table

Gage number	Diameter (mils)	Area (cmil)	Resistance (Ω/1000 ft)
1	289.3	83,690	0.1264
2	257.6	66,370	0.1593
3	229.4	52,640	0.2009
4	204.3	41,740	0.2533
5	181.9	33,100	0.3195
6	162.0	26,250	0.4028
7	144.3	20,820	0.5080
8	128.5	16,510	0.6405
9	114.4	13,090	0.8077
10	101.9	10,380	1.018
11	90.74	8,234	1.284
12	80.81	6,530	1.619
13	71.96	5,178	2.042
14	64.08	4,107	2.575
15	57.07	3,257	3.247
16	50.82	2,583	4.094
17	45.26	2,048	5.163
18	40.30	1,624	6.510
19	35.89	1,288	8.210
20	31.96	1,022	10.35
21	28.46	810.1	13.05
22	25.35	642.4	16.46
23	22.57	509.5	20.76
24	20.10	404.0	26.17
25	17.90	320.4	33.00
26	15.94	254.1	41.62
27	14.20	201.5	52.48
28	12.64	159.8	66.17
29	11.26	126.7	83.44
30	10.03	100.5	105.2
31	8.928	79.70	132.7
32	7.950	63.21	167.3
33	7.080	50.13	211.0
34	6.305	39.75	266.0
35	5.615	31.52	335.0
36	5.000	25.00	423.0
37	4.453	19.83	533.4
38	3.965	15.72	672.6
39	3.531	12.47	848.1
40	3.145	9.88	1069

*For copper wire at 25°C, average room temperature.

$$R_t = 1.619 + 1.619 \times (0.004 \times 50) = 1.943 \qquad (1.3)$$

The resistance has increased by 20 percent. If the wire is already loaded at capacity, this increased resistance will raise the temperature of the wire even more, and the resistance will further increase, and so forth.

Another effect of temperature is that it causes metal to expand and contract, especially at fittings and connections. It is a common perception that aluminum, for example, has a high thermal expansion coefficient. As the reasoning goes, aluminum expands and contracts more than copper, so it has a tendency to work loose from fittings that are not made up tightly. So, even though aluminum is cheaper and lighter than copper, it is not a popular material for electrical wiring.

A quick look at Table 1.1 shows that the thermal expansion of aluminum is actually the same as copper. However, aluminum has a greater resistance, so a larger wire is needed to carry a given amount of current. This larger size wire, then, will exhibit a greater sensitivity to temperature extremes and thus will expand more, and this causes the problems noted earlier. So the consequences of using aluminum wiring are as noted, although for a little more complex reason. Table 1.3 gives a comparison between copper and aluminum wire sizes.

Galvanic corrosion is another phenomenon to be wary of in electrical wiring systems. When two dissimilar metals are joined together, such as at a fitting, a small potential current is generated that over time can cause the materials to deteriorate. This is not a problem in low-voltage applications. Still, it is a good policy to try to specify that all el-

TABLE 1.3 Aluminum Cable Substitution

| | | Raceway size required for aluminum conductors | | | | | | | | |
| | | Number of conductors | | | | | | | | |
Copper size	Aluminum size	1	2	3	4	5	6	7	8	9
6	4	$\frac{3}{4}$	$1\frac{1}{4}$	$1\frac{1}{4}$	$1\frac{1}{2}$	$1\frac{1}{2}$	2	2	2	$2\frac{1}{2}$
4	2	$\frac{3}{4}$	$1\frac{1}{4}$	$1\frac{1}{4}$	2	2	2	$2\frac{1}{2}$	$2\frac{1}{2}$	$2\frac{1}{2}$
3	1	$\frac{3}{4}$	$1\frac{1}{4}$	$1\frac{1}{2}$	2	$2\frac{1}{2}$	$2\frac{1}{2}$	$2\frac{1}{2}$	3	3
2	1/0	1	$1\frac{1}{2}$	2	2	$2\frac{1}{2}$	$2\frac{1}{2}$	3	3	3
1	2/0	1	2	2	$2\frac{1}{2}$	$2\frac{1}{2}$	3	3	3	$3\frac{1}{2}$
1/0	3/0	1	2	2	$2\frac{1}{2}$	3	3	3	$3\frac{1}{2}$	$3\frac{1}{2}$
2/0	4/0	$1\frac{1}{4}$	2	$2\frac{1}{2}$	3	3	3	$3\frac{1}{2}$	4	4
3/0	250 MCM	$1\frac{1}{4}$	$2\frac{1}{2}$	$2\frac{1}{2}$	3	3	$3\frac{1}{2}$	4	5	5
4/0	300 MCM	$1\frac{1}{4}$	$2\frac{1}{2}$	$2\frac{1}{2}$	3	$3\frac{1}{2}$	4	4	5	5
250 MCM	400 MCM	$1\frac{1}{2}$	3	3	$3\frac{1}{2}$	4	4	5	5	5
300 MCM	500 MCM	$1\frac{1}{2}$	3	3	$3\frac{1}{2}$	4	5	5	5	6
350 MCM	500 MCM	$1\frac{1}{2}$	3	3	$3\frac{1}{2}$	4	5	5	5	6
400 MCM	600 MCM	2	$3\frac{1}{2}$	$3\frac{1}{2}$	4	5	5	6	6	6
500 MCM	750 MCM	2	$3\frac{1}{2}$	$3\frac{1}{2}$	5	5	5	6	6	

Notes:
1. All the conductors in any one raceway shall be either copper or aluminum.
2. Aluminum cable shall not be used for system or equipment grounds.
3. Do not use aluminum for wiring smaller than #4 AWG.

ements of a circuit should be of the same metal, including fittings, outlets, breakers, and all safety elements. This will minimize problems throughout the working lifetime of the power distribution system.

Voltage Drop

Notice from the tables above that the resistance of wire is constant. It is dependent only on the size composition of the wire. The power dissipated in the wire is therefore also a constant:

$$\text{Power loss} = i^2 R \sim \text{heat} \tag{1.4}$$

or "I-squared R" losses. This is power that is available at the source but that does not make it to the point of use because it is dissipated as heat. This is a major source of inefficiency in the electrical system. It increases by the square of the current. That means when the current increases by a factor of 3, the losses increase by a factor of 3^2, or nine.

The actual voltage drop in a specific circuit is easily calculated. In a two- or three-wire single-phase circuit the voltage drop is

$$V = \frac{(2K \times L \times I)}{D} \tag{1.5}$$

In a three-wire three-phase circuit the function is

$$V = \frac{2K \times L \times I}{D} \times 0.866 \tag{1.6}$$

In a four-wire three-phase circuit the function is

$$V = \frac{2K \times L \times D}{D} \times 0.5 \tag{1.7}$$

where V = drop in the circuit voltage
I = current in the conductor(s), A
L = one-way length of the circuit, ft
D = cross section area of the conductor, cmil
K = resistivity of the conductor

The factor K has the following values:

K = 11 for circuits carrying less than 50 percent of capacity (copper)
K = 12 for circuits carrying more than 50 percent of capacity (copper)
K = 18 for aluminum conductors

When performing these calculations, remember that the current-carrying capacity of the wire is decreased when multiple conductors or home runs are situated in a single conduit. These are governed by the following criteria:

Number of current-carrying conductors	Adjustment factor
4–6	80
7–9	70
10–20	50
21–30	45
31–40	40
41+	35

Once the allowable current is determined (adjusted for ambient temperature as necessary), that value is used for the allowable voltage drop calculations.

Resistance Losses

Resistance losses have several effects on the circuit. Most significantly, the voltage available at the far end of the circuit is diminished. This, in turn, implies that

- Lights burn dimmer.
- Motors run slower.
- Sensors are less exact.

This is especially true on circuits loaded close to the maximum, as is common with motors and lighting.

In the fluid circuit analogy, this means that the same current flows throughout the circuit (water flow rate is steady), but the head of pressure is less at the end because of friction in the pipe. The mechanical solution is to increase the diameter of the piping and to oversize the pump so that even the most distant element in the fluid circuit gets enough water, and at sufficient pressure. Increasing the pipe size is the best solution because this reduces the current, which by function is similar to Eq. 1.4. This reduces losses, often to the point of being able to downsize the pump. Increasing the size of the pump, on the other hand, increases the rate of flow and thus the losses. The circuit is supercharged so that despite the higher losses en route to the most distant point, there remains enough pressure to operate the fluid circuit element.

The larger diameter piping, though, can cost substantially more to install because of pipe supports, fittings, and other installation factors. So, even though the long-term energy use is greatly increased (especially because the pump has very long run hours), the solution of choice for mechanical designers is the larger-size pump. The solution is a little less clear cut for electrical designers.

The electrical analogy to this solution is to either increase the size of the service or to increase the conductor size. Increasing service capacity (i.e., voltage) is not as easy or as inexpensive as increasing pump size, and increasing the wire size is cheaper than increasing pipe size because often the conduit size is not increased. Also different combinations of wire sizes can be used to supply the same current, the smaller sizes being less expensive and easier to install. For these reasons it is often fairly economical to increase the current-carrying capacity of the wire to reduce resistance losses in the electrical distribution system. This, of course, avoids the fact that there are only a few standard voltages available to the electrical designer, and an unlimited range of pump pressures are available to the mechanical designer.

There is another benefit that tips the scales of the electrical analysis in favor of modifying wire sizes versus source voltage. Just as higher flow rates through hydronic systems from an oversized pump increase thermal losses from the fluid (because the flow is more turbulent), so too with electrical systems. The thermal loss, in this instance caused by the flow resistance of the wire, causes a higher heat loss into the conditioned space. This is a penalty in the cooling season because the heating, ventilation, and air conditioning (hvac) system must remove the heat. The mechanical analogy does not have this flaw. Even though this inefficiency is not charged against the electrical system, it can be a sizable portion of the hvac load.

Finally, the efficiency of the mechanical system decreases with time as years of fluid flow pits the inside of the pipe and mineral deposits cause the inner surface to become rough and irregular. This increases the pressure losses, and the pump must work harder to supply the needs of the loop. Electrical systems have a similar problem, as the heat from resistance losses causes the insulation to break down. This, in turn, allows more heat to escape. Mechanical designers anticipate this drop in performance by originally designing the pipe based on old, rough metal. Electrical designers do not usually make any such allowances.

Circuit Elements

The actual wire is not the only part of a circuit that causes resistance losses. Just as each valve or fitting in a hydronic system causes turbulence and pressure losses, so too with electric circuit elements that interrupt the uniform length of current-carrying wire. Some examples are

- Lugs and connectors in junction boxes, wireways, and panels
- Fuses, circuit breakers, and disconnects
- End connectors at outlets, switches, and other control devices

Whereas a fluid system loses energy in the form of vibration, heat transfer, and reduced pressure, the electrical losses are all manifested as heat. That is why electric closets always need a significant amount of cooling by the hvac system all year around.

Mechanical designers are careful to minimize control elements and fittings in the mains—such as reducers and balancing valves—and when such items are necessary, they are provided with a full bore to minimize pressure losses. The reason for this is that every restriction in the fluid path adds an additional steady load on the pump. Not only must the pump size be increased to handle this extra static head, but also the operating costs increase.

Electrical designers are seldom so diligent, even though the same principles apply. This is because the electric panel and service do not have to be as carefully designed as their mechanical counterparts. The voltage and amperage are provided only at specific levels. In both cases, systems are designed to the next higher level, so there is always plenty of extra capacity available. Pumps, piping sizes, and hydronic control elements are often reduced in the value-engineering phase of projects. Electrical panels, conduit sizes, and routing of feeders are rarely so closely scrutinized, even though just as much benefit can result—if not more, because element for element the electric system is more costly.

There are a number of things that can be done to reduce the electrical losses in the main distribution grid. In most cases these modifications also reduce the project cost and potential for fire hazard as well.

- Minimize the use of connecting elements and splices in favor of long, continuous runs of wire.
- Locate heat-producing elements outside of the conditioned space but still in an adequately ventilated space.
- In all cases make provisions to extract the heat produced; otherwise the equipment will be even less efficient.
- Instead of having two separate control elements, combine protection and connection means by such practices as using a fused disconnect switch in lieu of a disconnect with a separate fused element.
- Where economically feasible oversize connecting elements, such as wire in a multifeeder wireway, to minimize resistance losses.
- Use connecting means (e.g., bolts and lugs) that have the maximum contact area (the smaller this area, the greater the local resistance and the more heat generated there), and use all that contact area in the actual connection.
- Secure all connections tightly, especially where multistranded wire is used (i.e., ensure all strands are bound in the connection).

Most of these items are a simple matter of diligence and planning.

Good work habits ensure firm connections and direct routing of conduits with a minimum of fittings. Thorough planning helps reduce the number of fittings, and good design minimizes the number of safety and control elements. Together these practices can reduce the cost of the work by reducing the length of conduit runs, the number of fittings, and the size and capacity of protecting means, and can also result in a more efficient installation.

Insulators

Insulating materials are usually not given as much attention as conductors. Some common insulators in order of decreasing conductivity are

Slate

Oils

Porcelain

Dry paper

Silk

Sealing wax

Ebonite

Mica

Glass

Dry air

Many of these materials are in common use for electrical applications. Porcelain insulators are used for overhead electrical wires, where they are supported on poles or at the service entrance to a building. Notice that glass is a better insulator and would be preferred in such applications, except it is easier to break and not as durable. Oil-filled transformers are common, despite oil having a higher conductivity than air in a dry-type transformer. The oil is used, among other reasons, to help transfer heat from the coil to the casing. Air is a thermal insulator too, so it cannot transport heat as easily from the transformer core. Dry paper is a good insulator, especially if impregnated with a wax. This combination can be found in motors to insulate the windings or in cable to insulate between individual plastic coated conductors.

Plastic Insulated Sheathed Cables

Residential and commercial wiring consists of insulated cables, typically with a durable outer plastic or metal-clad jacket. The wires inside the cable are insulated with a type of polyvinyl chloride (PVC).

Many older buildings have two conductor cables. All new work must be done with cable that has an integral ground wire. Some cable is stenciled "W/G" to denote "with ground." Several common designations are important to commercial work.

Nonmetallic sheathed cable

Type nonmetallic (NM) can be used inside only, and only in dry locations. Type nonmetallic cable (NMC) is acceptable to damp locations, indoors. Both types are limited by code to three-story buildings, and with a purpose other than assembly. There are some instances where NM and NMC can be embedded in masonry or plaster, but generally they must be in accessible locations.

Type UF (for underground feeder) cable is suitable for outdoor use, underground, and in wet locations. Outdoor circuits do not always need to be grounded, although it is best to have integral grounding in all applications. It is allowable to direct-bury Type UF cable, unless there is a possibility of mechanical damage, in which case the cable should be run in conduit. If there is any doubt, it is always best to install the cable in gray electrical PVC conduit. Always mark the conduit run with metal markers that can be detected by a surface metal detector. Type UF cable cannot be used for a service entrance. It is also prohibited in areas of assembly, studios, and embedded in concrete. These are specialty cable applications. Table 1.4 provides guidance on the depth of cover required for direct burial cable carrying 0- to 600-V conductors.

Type SE (for service entrance) and Type USE (for underground service entrance) are acceptable for use in service entrance installations. Generally, it is best to have an insulated ground with this type of cable, although it is mandatory when used as an interior feeder or branch circuit, as is often the case. The USE cable is rated for underground use. Again, it is best for the ground to be insulated. Either type, when comprised of two or more conductors, is permitted to have one uninsulated conductor when used as service entrance conductors. However, this is not advisable for safety purposes and for labeling.

The most common insulation for commercial applications is THHN. It is suitable for wet and dry locations, having a composition of moisture and heat resistant thermoplastic. Other insulation types are listed in Table 1.5.

Feeder Size Calculations

Table 1.6 can be used to select wire size for a power circuit at various voltage drops. First, the circuit wattage is determined using the following guidelines:

TABLE 1.4 Minimum Cover Requirements, 0 to 600 V, Nominal*

Location	Direct burial	Rigid metal conduit or IMC	Rigid nonmetallic conduit†	Residential branch ckts 120v or less‡	Specialty circuits¶
All others	24	6	18	12	6
See note 1	18	6	12	6	6
See note 2	0 (raceway)	0	0	0 (raceway)	0 (raceway)
See note 3	18	4	4	6 (direct) 4 (raceway)	6 (direct) 4 (raceway)
See note 4	24	24	24	24	24
See note 5	18	18	18	12	18
See note 6	18	18	18	18	18
See note 7	2 (raceway)	2	2	2 (raceway)	2 (raceway)

*Cover is defined as the shortest distance measured between a point on the top surface of any direct buried conductor, cable, or other raceway and the top surface of finished grade, concrete, or similar cover.
†Approved for direct burial without concrete encasement or other approved raceways.
‡With GFCI protection and maximum over current protection of 20 A.
¶Circuits for control of irrigation and landscape lighting limited to not more than 30 V and installed with Type UF or in other identified cable or raceway.
Notes:
1. In trench below 2-in-thick concrete or equivalent.
2. Under a building.
3. Under minimum of 4-in-thick concrete exterior slab with no vehicular traffic and the slab exceeding not less than 6 in beyond the underground construction.
4. Under streets, highways, roads, alleys, driveways, and parking lots.
5. One- and two-family dwelling driveways and outdoor parking areas, and used only for dwelling-related purposes.
6. In or under airport runways, including adjacent areas where trespassing is prohibited.
7. In solid rock where covered by minimum of 2 in of concrete extending down to rock.
SOURCE: Reprinted with permission of NFPA 70–1993, National Electric Code © 1993.

- Use actual wattage for large loads such as ranges, microwaves, water heaters, and refrigerators. It is usually best to have a dedicated circuit for such major appliances.
- Assume 250 W for each duplex outlet.

The same table can be used to calculate the feeder size for the entire panel. Once the panel schedule is complete and the existing load is totaled, add 500 W for each spare circuit, and add 20 to 50 percent for future expansion.

It is always important to keep an accurate record of all calculations made on a project. This includes neat tables or forms of all the voltage drop and sizing calculations that were performed as a part of the project design phase. Most designers do all the calculations in their head, on scraps of paper, or not at all. It takes very little more time to keep a formal record of the work done, to include all assumptions made such as wire type, operating temperature, and number of conductors

TABLE 1.5 Characteristics of Insulated Conductors

Trade name*	Type letter	Max. temp.	Application provisions
Asbestos (*no longer used*) heat-resistant, fixture wire	AF	150°C 302°F	Fixture wiring Limited to 300 V and indoor dry location
Fluorinated etheylene propylene	FEP	90°C 194°F	Dry and damp locations
	FEPB	200°C 392°F	Dry locations (special applications)
Perfluoroalkoxy	PFA	90°C 194°F	Dry and damp locations
		200°C 392°F	Dry locations (special applications)
Perfluoroakloxy	PFAH	250°C 482°F	Dry locations only; only for leads within apparatus or within raceways connected to apparatus
Silicone rubber insulated fixture wire	SF-1	200°C 392°F	Fixture wiring Limited to 300 V
Solid or seven-strand	SF-2	200°C 392°F	Fixture wiring as permitted in NEC Section 310-8
Heat-resistant synthetic polymer	RH	75°C 167°F	Dry locations
Moisture and heat-resistant rubber	RHH	90°C 194°F	Dry locations
Heat-resistant synthetic polymer	RHW	75°C 167°F	Dry and wet locations. For over 2000 V, insulation shall be ozone resistant
Heat-resistant synthetic polymer	RHW-2	90°C 194°F	Dry and wet locations
Thermoplastic	T	60°C 140°F	Dry locations
Thermoplastic and fibrous outer braid	TBS	90°C 194°F	Switchboard wiring only
Extended polytetrafluoroethylene	TFE	250°C 482°F	Dry locations only; only for leads within apparatus or within raceways connected to apparatus
Moisture-resistant thermoplastic	TW	60°C 140°F	Dry and wet locations
Heat-resistant thermoplastic	THHN	90°C 194°F	Dry locations
Moisture and heat-resistant Thermoplastic	THW	75°C 167°F	Dry and wet locations
Underground feeder	UF	60°C 140°F	Moisture resistant
Underground service entrance cable	USE	75°C 167°F	Heat and moisture resistant

TABLE 1.5 Characteristics of Insulated Conductors (*Continued*)

Trade name*	Type letter	Max. temp.	Application provisions
Heat-resistant cross-linked synthetic polymer	XHH	60°C 194°F	Dry and damp locations
Moisture and heat-resistant cross-linked thermosetting polyethylene	XHHW	90°C 195°F 75°C 167°F	Dry locations Wet locations
Thermoplastic and asbestos (*no longer used*)	TA	90°C 194°F	Switchboard wiring only
Mineral insulation (metal sheathed)	MI	85°C 185°F 250°C 482°F	Dry and wet locations with Type O termination fittings For special applications
Moisture-, heat-, and oil-resistant thermoplastic	MTW	60°C 140°F 90°C 194°F	Machine tool wiring in wet locations Machine tool wiring in dry locations
Silicone-Asbestos	SA	90°C 194°F 125°C 257°F	Dry locations For special applications
Varnished cambric	V	85°C 185°F	Dry locations only; smaller than No. 6 by special permission only
Asbestos and varnished cambric	AVA	110°C 230°F	Dry locations only
Asbestos and varnished cambric	AVL	110°C 230°F	Dry and wet locations
Asbestos and varnished cambric paper	AVB	90°C 194°F 85°C 185°F	Dry locations only For underground service conductors or by special permission
Rubber-covered fixture wire	RF-1	60°C 140°F	Fixture wiring Limited to 300 V
Solid or 7-strand	RF-2	60°C 140°F	Fixture wiring and as permitted by NEC Section 310-8
Heat-resistant rubber covered fixture wire	RFH-1	75°C 167°F	Fixture wiring Limited to 300 V
Solid or 7-strand	RFH-2	75°C 167°F	Fixture wiring, and as permitted by NEC Section 310-8
Thermoplastic-covered fixture wire	TF	60°C 140°F	Fixture wiring, and as permitted by NEC Section 310-8
Modified ethylene tetrafluoroethylene	Z	90°C 194°F 150°C 302°F	Dry and damp locations Dry locations (special applications)

TABLE 1.5 Characteristics of Insulated Conductors (*Continued*)

Trade name*	Type letter	Max. temp.	Application provisions
Modified ethylene tetrafluoroethylene	ZW	75°C 167°F	Wet locations
		90°C 194°F	Dry and damp locations
		150°C 304°F	Dry locations (special applications)
Heat-resistant, thermoplastic covered fixture wire	TFN	90°C	Fixture wiring, and as permitted by NEC Section 310-8

*Some of these wire types are no longer manufactured but are provided here for identification purposes. *Note especially wires containing asbestos, which are rarely considered in abatement projects.*
SOURCE: Reprinted with permission of NFPA 70–1993, National Electric Code © 1993.
NEC is the registered trademark of the NFPA, National Fire Protection Association, Quincy MA 62269. This material is not the complete and official position of the NFPA on the referenced subject, which is represented by the standard in its entirety.

in a conduit. Liability is the most compelling reason to keep such formal records in a separate project folder or three-ring binder. If liability issues arise as a result of a bad installation or other cause beyond the designer's control, written proof of calculations will go a long way in avoiding any blame.

TABLE 1.6 Voltage Drop on DC or 100 Percent Power Factor AC Circuits*

Wire size	1	2	3	4	5	6	7	8
14	177	354	531	705	866	1,060	1,060	1,240
12	282	563	845	1,130	1,410	1,690	1,970	2,250
10	448	895	1,340	1,790	2,240	2,690	3,130	3,580
8	712	1,420	2,140	2,850	3,560	4,270	4,980	5,690
6	1,110	2,220	3,330	4,440	5,550	6,660	7,770	8,880
4	1,760	3,530	5,290	7,060	8,820	10,600	12,300	14,100
2	2,880	5,750	8,630	11,500	14,400	17,300	20,100	23,000
1	3,540	7,070	10,600	14,100	17,700	21,200	24,800	28,300
1/0	4,460	8,910	13,400	17,800	22,300	26,700	31,200	35,600
2/0	5,620	11,200	16,900	22,500	28,100	33,700	39,300	44,900
3/0	7,080	14,200	21,200	28,300	35,400	42,500	49,600	56,600
4/0	8,920	17,800	26,700	35,700	44,600	53,500	62,400	71,300
250 MCM	10,500	21,000	31,600	42,100	52,600	63,100	73,700	84,200
300 MCM	12,600	25,200	37,800	50,400	63,000	75,600	88,200	101,000
350 MCM	14,700	29,300	44,000	58,700	73,300	88,000	103,000	117,000
400 MCM	16,700	33,400	50,200	66,900	83,600	100,000	117,000	134,000
500 MCM	20,800	41,500	62,300	83,100	104,000	125,000	145,000	166,000
750 MCM	30,500	60,900	91,400	122,000	152,000	183,000	213,000	244,000
1000 MCM	39,500	79,000	118,000	158,000	198,000	237,000	277,000	316,000

*Find the ampere · feet by multiplying current in amperes by length of one wire in feet (not the total length of wire in the circuit).

Chapter 2

Electrical Circuits

This section reviews some of the basic concepts of electricity and its use and propagation. Electrical designers without any formal education in the field may be familiar with only the application of the principles. Degreed engineers may have forgotten many of the ideas, because they are rarely used explicitly in their most basic form. Instead, applications engineers are accustomed to using charts, tables, or computer programs, all of which have the fundamental equations built in but are not in sight.

There are times when a calculator is not available and you must do arithmetic in your head; so too with electrical analyses. A good sense of numbers and relative values will resolve many problems, just like a good understanding of electricity benefits an electrical contractor or designer.

This section deals with direct current (dc) circuits. The math is much easier than with alternating current (ac) circuits. The operating principles are exactly the same, however, so a study of dc characteristics fosters the fundamental comprehension that is sought.

Ohm's Law

The most important relationship in electrical design is called Ohm's Law. It defines the relationship between current and voltage in a simple circuit.

$$I = \frac{V}{R} \tag{2.1}$$

where I = current, A
V = potential, V
R = resistance, Ω

Notice that the constant of proportionality is nothing more than a value that identifies the way that the current varies with respect to the voltage. It is an intrinsic, linear characteristic of the circuit in question.

Ohm's Law is a very compact phrase that says a lot about what happens in an electrical circuit. (As accustomed as the reader may be to dealing with alternating current, pause now and then to consider each idea in terms of ac circuits you are familiar with. There is, you will find, virtually no difference from dc circuits.) Imagine a light bulb in a circuit that can be attached to various voltage terminals or batteries in series. As each 2-V battery is added to the circuit, the voltage in the circuit increases by 2 V (Fig. 2.1).

Although you may not be accustomed to thinking of it as such, an incandescent light bulb is really just a high-resistance element in a vacuum-tight glass enclosure. As current flows through the filament, a high resistance dissipates the current as heat—in this case, sufficient heat to make the element glow bright yellow. Much of the heat created is therefore dissipated as light. Another simple function can be used to determine just what the resistance of the carbon filament is. That is, the identity of electrical power is defined as

$$P = I^2 \times R = \left(\frac{V}{R}\right)^2 \times R = V^2 \times R \qquad (2.2)$$

where P = power, W
I = current, A
R = resistance, Ω

So, to find the resistance of the incandescent lamp, assume a 100-W bulb in a 120-V circuit and substitute in Eq. 2.1:

$$100 = \frac{120^2}{R}$$

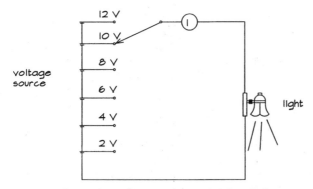

Figure 2.1 Increasing voltage produces a brighter light.

$$R = 144 \, \Omega$$

This example illustrates what happens in any circuit with a resistance (usually wire sizes are calculated on a 2 percent voltage drop, caused by wire resistance). Power is dissipated by the resistance as heat, and sometimes as light in the case of short circuits and even some engineered elements, such as starter terminals or circuit breakers.

Fuses operate by this principle. As current flows above the rated value, the metal element in the fuse heats up. If the current is high enough and lasts long enough, the fuse element will melt and open the circuit.

Current in a Series Circuit

The next important concept is that the current in every element of a series circuit is the same (Fig. 2.2). There is only one path for the electrons to follow, so they all do the same thing. Water in a pipe loop with heat exchangers creating resistance elements is exactly analogous. The water has only one possible path, so it all goes the same way.

Figure 2.2 is called a series circuit. In such an in-line arrangement, the equivalent resistance of the two resistors is

$$R = R_1 + R_2 \qquad (2.3)$$

So if both resistors are 10 Ω, the total resistance that the voltage source sees is 20 Ω. If this was a 120-V circuit, the current flowing would be

$$I = \frac{V}{R} = \frac{120}{20} = 6 \, \text{A}$$

Double the imposed voltage, with the resistance unchanging (because it is independent of both voltage and current), and the current doubles.

What happens if one of the resistances is changed by a person? If a conducting wire is held in each hand, the resistance of the body is at least 10,000 Ω. The current flowing would then be approximately

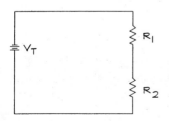

Figure 2.2 Series resistances.

$$I = \frac{120}{10{,}000} = .012\,\text{A} = 12\,\text{mA}$$

This can be fatal. As little as .01 mA can cause an electric shock.

Notice that the power dissipated in the body by the original voltage and current is

$$P - I^2 \times R = (.012)^2 \times 10{,}000 = 1.44\,\text{W}$$

If the source current is unable to supply this power, the current drops to the current that the available power can supply. The greatest danger, then, is from a source over 30 V that has enough power to maintain the load current through the body.

Parallel Circuits

Referring again to Ohm's Law, note that the voltage drop is divided across the resistances in the series circuit in proportion to their value by the function

$$V_1 = I \times R_1 \quad \text{or} \quad V_2 = I \times R_2 \tag{2.4}$$

where the sum of the voltage drops through all the elements of the series circuit are equal to the potential difference applied to the circuit by the battery.

Another characteristic of series circuits is the effect of an open circuit in the current path. This is like replacing R_1 or R_2 with an open element (i.e., one that has an infinite resistance). The result is that no current flows in the whole loop, so there is no voltage across the remaining resistance in the circuit. This is the same as though a valve was completely closed in a fluid circuit, causing all flow to stop.

A short circuit, on the other hand, is when a resistor is removed and one of the open terminals is grounded. This creates a path of zero resistance. So the current flows without limits, other than the capacity of the power source and the remaining resistance in the circuit. By the fluid analogy, the valve in a fluid circuit is wide open and the entire volume of flow exits the pipe at that location. The example of a person with the wire in either hand resembles the short circuit most closely, so the person is virtually a ground for the current.

Parallel Circuits

A second common circuit arrangement is called a parallel circuit. Two items are connected across the poles of a battery, or the phases of a circuit, as illustrated in Fig. 2.3. The first thing to realize about par-

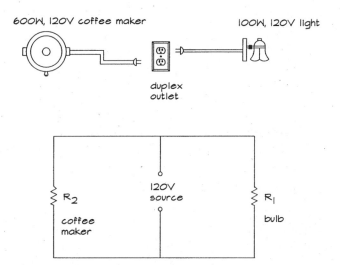

Figure 2.3 Appliances connected in parallel.

allel circuits is that the same voltage is applied across both resistances. This, in turn, implies that the two parts of the circuit have different values of current. The total current that flows from the battery, in this example, is equal to the total of the current that goes through the bulb plus the current in the heating element. The mainline current equals the sum of the branch-line currents.

The resistances in circuits are added a little differently in parallel circuits. The total resistance, for example, of the elements shown in Fig. 2.3 is the reciprocal of the sum of the reciprocals:

$$\frac{1}{R_T} = \frac{1}{R_1} + \frac{1}{R_2} + \frac{1}{R_3} \qquad (2.5)$$

This means that a circuit can replace the circuit of Fig. 2.3 with just one equipment resistance, R_T, in series with the voltage source.

An open circuit in a parallel circuit simply stops current delivered to that branch. All other components of the circuit remain powered as before. The lights in most ornamental tree strings are wired in parallel. When one light burns out, the others are not affected. Some strings, however, are series wired. When one bulb burns out or is removed, they all go out.

A short circuit is another matter altogether. As illustrated in Fig. 2.4, a short across one of the parallel branches shorts them all. Because the short path has practically zero resistance, the current flowing is unchecked. Instead of the rated 1A, it could be 100 A. There are two possible dangers from the short circuit. First, the source is

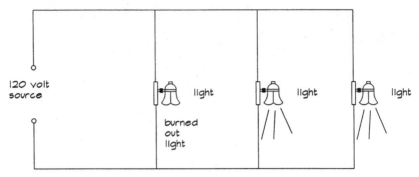

Figure 2.4 Parallel circuit short circuiting.

not designed to maintain much more than its rated output. Doing so can cause it to heat up or even burn up. The same applies to the wires carrying the high current. As the wire heats up, the insulation can burn and ignite a fire that spreads through the building.

The purpose of fuses and circuit breakers is to stop this from happening by interrupting the circuit before the current gets too high. A sensitive element in the fuse melts before any other damage can occur and the branch circuit is disabled.

Series-Parallel Circuits

Many common circuits are composed of a combination of parallel and series branches. An example is Fig. 2.5. This circuit is similar to a four-lamp fluorescent fixture with two ballasts (not shown for clarity), two lamps circuited from each. When one lamp in either pair goes out, the other does too. These lamps are wired in series. The other two

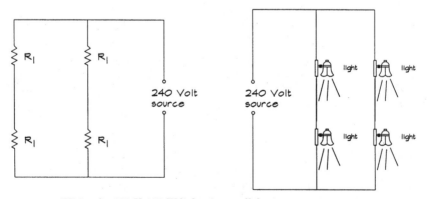

Figure 2.5 Wiring for 120-V, 100-W lights in parallel.

lamps in the fixture are not affected by this outage. They are wired in parallel to the other two lamps. Some other characteristics are as follows:

- The current in each branch equals the voltage applied divided by the total resistance.
- The voltage in each branch is divided between the two resistances in series. If the voltage is the same, there is 120 V across each bulb. The same current flows through both bulbs.

Whetstone Bridges

There are many controls applications that are based on resistance. Thermisters and resistance thermometers measure temperature, and strain gages measure pressure; both do so by virtue of changes that are directly proportional to resistance of the circuit. All that needs to be done is to integrate the device into a special circuit that can be balanced to determine the change in resistance of the unknown element. Figure 2.6 shows a simple whetstone bridge circuit that is often used for this purpose.

A main voltage is applied across two of the terminals, and the differential in current flowing in the other two branches is measured. The bridge is balanced (i.e., the meter reading is null) when the ratio of resistance of any two adjacent arms is equal to the ratio of resistances in the other two arms, taken in the same sense.

There are several possible configurations shown in Fig. 2.7. The unknown resistance (i.e., a remote thermister or strain gage to measure pressure) is in one of the branches. Another branch is adjustable, by one of the methods indicated (i.e., rheostats or variable resistances). When the adjustment zeros the meter reading, the unknown resistance can be calculated using the rules of parallel and series circuits.

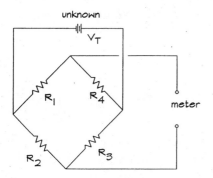

Figure 2.6 Whetstone bridge circuit.

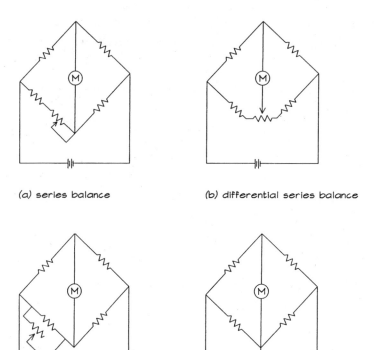

Figure 2.7 Balancing dc resistance bridges.

Whetstone bridges are quite common, especially in controls systems for hvac equipment. The components are inexpensive, dependable, and easily repaired. They can also measure a very wide range of values if the circuit voltage and fixed resistance values are carefully selected.

Adjusting Bridge Sensitivity

One important characteristic for bridge circuits in many controls applications is the ability to adjust the sensitivity of the device. There are several reasons for this capability:

- To attenuate inputs that are too large
- To calculate the input to the system to be aligned with the particular scale on the readout instrument

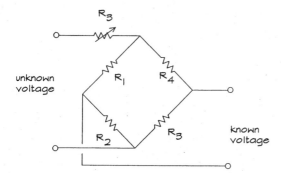

Figure 2.8 Adjusting bridge sensitivity with a variable resistance.

- To give the system flexibility to be used with many different tranducers by having the ability to be configured or customized to each device after it is installed in the circuit
- To control extraneous inputs, such as ambient (space) temperature or nominal system pressure, so that the net output is the change in temperature or pressure

A simple way to adjust the output of a whetstone bridge is shown in Fig. 2.8. A variable series resistance is placed in one (or both) of the leads to the input. To see its effect, assume all the resistance elements R have the same value. The resistance seen by the voltage source will also be R. Inserting the series resistance as shown makes it into a voltage divider and reduces the input to the bridge itself by the factor

$$n = \frac{R}{R + R_S} = \frac{1}{1 + \left(\frac{R_S}{R}\right)} \quad (2.6)$$

where n is the bridge factor. The bridge output is reduced by a known, proportional amount, so the effective sensitivity is reduced somewhat. Thus, if the original bridge is ultrasensitive, the adjustment of the input resistance will make it useful with a wide range of sensors so that all can be adjusted to match a common calibration scheme.

Kirchoff's Laws

There are many kinds of circuits that cannot be analyzed as series, parallel, or series-parallel configurations. Examples are an unbalanced bridge circuit or a circuit in which two voltages are applied in different branch circuits. The latter is common in controls applications, when line voltage is stepped down to a lower controls voltage,

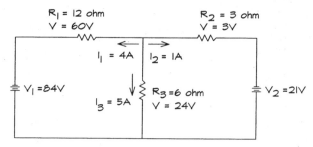

Figure 2.9 Circuit analysis.

such as 12 or 6 V. The general method of solving such applications is to use the series and parallel circuit methods to reduce the complexity of the circuits to the extent possible and then to apply other methods. As illustrated in Fig. 2.9, the circuit is not susceptible to the rules developed thus far.

Kirchoff's Current Law

This simple relationship evaluates the electric currents entering and leaving a single node (where several branches intersect). The sum of currents entering the node must equal the sum of currents leaving the node. This is just as you would expect in a fluid loop. Otherwise it would not be a steady-state system.

As an example, consider node c. It is the intersection of three separate elements of the circuit, each with different currents. If all currents are assumed to enter the node, the equation of Kirchoff's Current Law must be satisfied:

$$I_1 + I_2 + I_3 = 0 \qquad (2.7)$$

Upon analysis, one or more of these currents will be negative, which simply means the direction is opposite to that assumed here.

Kirchoff's Voltage Law

The next rule for closed circuits is that the sum of voltages around any closed path is equal to zero. There are three such closed loops in Fig. 2.9, one around the entire perimeter and one around each cell. Notice that adding up voltages around each of these closed paths equals zero.

This is illustrated in a fluid analogy, too. A pump supplies a specific pressure to a closed loop. This pressure is used up at valves, fittings, air handler coils, and other devices that cause a pressure drop due to friction losses. If the total resistance is lessened, the flow increases according to the pump's capacity.

The solution method is to use Kirchoff's Voltage Law to create loop equations for each of the closed paths. For example,

$$V_T - V_1 - V_2 - V_3 = 0 \qquad (2.8)$$

where IR can be substituted for each V that is unknown. Two more such equations can be created for the other two closed loops in the cells.

Other functions can be generated by Kirchoff's Current Law for the branch currents and their algebraic relationship at the two nodes of the circuit. These currents are the same as branch currents used in the three voltage equations.

This gives a set of five independent equations that can be used to solve for the five unknown voltages. The three currents are not strictly independent unknowns because there is a relationship between an unknown voltage and a known resistance, so they are a derived function. The entire set of values can be solved using basic algebra.

Three-Phase Power and Neutral Current

All of the analyses so far have been for single-phase systems. It is popular to represent multiphase systems by the transformer connections. Before delving into three-phase systems, it is helpful to illustrate how single-phase current is developed from the transformer windings.

Figure 2.10 shows three common types of connections to the secondary side of a transformer. The transformer is configured to supply 120/240 V, single phase. In the first diagram, the device can deliver 120 V from one phase line and ground. In the middle illustration, the transformer has a midpoint (center) tap. Either phase line can be tapped to deliver 120 V, single phase, between the phase line and the neutral. At the same time, both phase lines can be used to supply 240 V, single phase. (Note that the two secondary windings have been connected together to put both windings in series, resulting in a neutral connection.)

In this example, all power is single phase because there are only two conductors. That is, voltage between the first and second conductors and between the second and third is single phase at 120 V, and be-

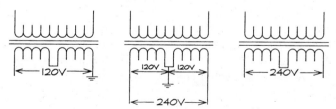

Figure 2.10 Single-phase transformer connection.

tween the first and third phase it is single phase at 240 V. These conductors do not change phase at the same time, so they are not in phase. If all three conductors are connected to the transformer shown, the system will operate in three phase. The conductors will be in phase.

Most modest-sized installations benefit from having both voltages available at single phase. Single-phase 120-V power is needed for outlets, lights, and small equipment loads. Water heaters, air conditioning condensing units, strip heat for air handlers, and other appliances use the 240-V single-phase power. Only larger commercial sites have the large motors that require three-phase power, for elevator motors, chillers, and large air handler fans.

Alternating Current

To this point the flow of current in dc circuits and how it behaves in simple resistance circuits have been described. This same simple current flows through the windings of a motor and induces the iron core to become a magnet with north and south poles. When there is a single set of poles, a single phase is created. Three sets of poles will generate three-phase current.

A periodic waveform is created by a loop of wire rotating in a magnetic field. As the loop rotates within the magnetic field, a voltage is induced across the ends of the loop shown. The amount of voltage induced in the wire is proportional to the number of flux lines the loop crosses. The strongest flux lines run straight between the poles as shown. As the loop moves a small angle through 90°, it cuts across many flux lines, moving in a small arc perpendicular to the lines. Moving through the same small angle at 180°, it cuts very few magnetic flux lines because it is moving almost parallel to them. Thus the induced voltage will be highest at 90° and lowest at 180°.

As the loop current makes one complete revolution, the voltage goes from zero to maximum, then back to zero. In the second half of the revolution, the loop has the opposite orientation with respect to the magnetic field, so the voltage follows the same pattern but in the opposite, negative sense. This pattern is called a sinusoidal wave, or sine wave for the sine of an angle (i.e., the magnitude of the opposite side versus the hypotenuse of a right triangle). As the angle of the loop with respect to an arbitrary starting point at zero moves through 360° of a circle, the value varies uniformly between zero and unity, or 1.

When this wave pattern is imposed on a circuit, a current flows in reaction to the changing voltage. The current will change directions in sync with the alternating pattern of the voltage.

Chapter 3

Phase Diagrams

The discussion thus far has been two-dimensional. This works in the discussion of resistors and single-phase current, and actually, it works for the design of most electrical devices. Three-phase systems get a little complicated, but a few basic equations and all the necessary voltages, amps, and other circuit parameters can be extrapolated from the system configuration. In fact, you can design complex electrical systems for a whole career, never knowing more than this 2-D perspective of power.

The problem is that in this 2-D world there is no true understanding of three-phase power and its implications. Therein lies the danger. If you don't know exactly what it is you are creating, designing, or installing, there is always the tendency to overdesign. When in doubt, bigger is better. However, electrical systems do not always work like that because safety requires adequate provisions be made to protect against the bigger voltage, current, or frequency. A 100-A circuit breaker is not better than a 30-A one when the protection of personnel and equipment requires the smaller device. A 65K AIC 100-A panel is not better than a 14K AIC 100-A panel in protecting an electrical system against the dire consequences of short circuits.

At least, such devices are not usually adequate. In those few situations where they are, it is important to have a firm grasp of the actual dynamics of power and its use in electromechanical devices. This is where it is important to be able to visualize electric current in three phases and its behavior in the most basic devices.

Sinusoidal Waveforms

The first task in the understanding of alternating current is to be able to visualize it graphically and mathematically. The actual pattern of alternating current is a sinusoidal wave, which is nothing more than

32 Chapter Three

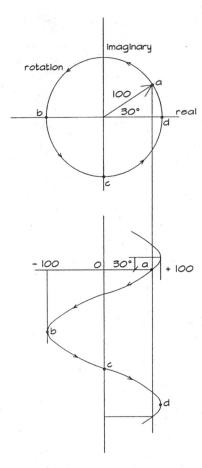

Figure 3.1 The complex plane and sinusoidal motion.

the projection of a circle onto a plane, as illustrated in Fig. 3.1. If the point on the circle rotates 60 times per second, it generates a sine wave that has a frequency of 60 Hz. The voltage varies from maximum to minimum and back with each rotation on the circle, or cycle. This constitutes one period, and the distance from one start point to the next is the wavelength. The maximum distance from the origin is the amplitude of the function, which might be 120 V.

It is important to recognize that the polarity of the wave changes with each cycle. Connect an oscilloscope to a conductor carrying such a wave, and the exact sine wave pattern will be created on the CRT. (Connection to a regular voltmeter will not fluctuate because the device measures the average amplitude of the voltage.)

To get a feel for the system of three-phase current, imagine three such patterns spaced one-third a wavelength apart. Picture these waves energizing a motor rotor, successive peaks striking the rotor in turn. As each wave peak passes, the effect by induction is to push the rotor a little; both positive and negative peaks act to push the rotor in a counterclockwise motion, like water on a water wheel. There are many sets of windings in a motor, each having a peak at a little later time than the previous set. The net effect on the rotor is a steady impulse in the clockwise direction.

If the sine wave designates the fluctuation of voltage in a circuit, the current is also a sinusoidal wave. However, because current is the rate of change of voltage, it is one-fourth of a wavelength (or 90° if you consider one wavelength is the projection of a single rotation around 360° of a circle) out of phase with the voltage. This phase relationship between voltage and current is maintained in all the circuits to be analyzed, even though the voltage or current through a specific device may vary from the sense of the charging fields.

Inductors

One circuit element has already been evaluated in great detail, the resistor. Now a dynamic element will be studied, the inductor. A coil of wire wrapped in tight loops acts like an inductor. The usual circuit symbol is a small coil, as shown in Fig. 3.2a. The accompanying graph in Fig. 3.2b shows how the circuit behaves when the switch is closed. The current in the inductor increases steadily until a steady state is reached and the current is the same as across the battery. This period of change in the circuit is called the transient response.

This circuit, you will notice, is shown excited by a direct current source. Excitation by an ac source will be discussed later. The mechanical analogy of the inductor is a mass acted upon by a force. The mass acts to oppose the force initially, until it reaches a constant velocity in a direct analogy to electrical inductance. This is just how the current moving through an inductor behaves.

The net action by the inductance is to store energy, as shown in Fig. 3.2c. In an ac field, the inductor is constantly charged and discharged, first storing energy from the circuit and then releasing the energy to the circuit in a uniform ebb and flow of energy. This is just what happens in the motor core. In fact, the multiple windings on a motor cause it to behave as an inductor, storing and discharging a portion of the electrical power used to energize it.

Capacitors

Two parallel flat plates, as illustrated in Fig. 3.3, symbolize a capacitor. When the switch is opened, electric charge moves to the plates

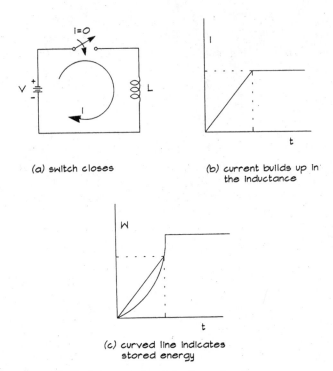

Figure 3.2 Performance of an inductance circuit.

and starts to build up on them. Current continues to flow until the plates have the same charge as the voltage source. At that time the current flow stops and, as shown in Fig. 3.3c, the voltage remains constant. The mechanical analogy to the capacitor is a spring. The initial force causes a steady rate of displacement of the spring until the force is exhausted and the spring can be stretched no more.

The capacitor stores energy as well, between the plates. Just as a spring recoils back to its original condition when the force is removed, the capacitor releases its charge as soon as the voltage is stopped. Excitation of a capacitor by an ac source has much the same response as an inductor. Energy is stored and released. The capacitor releases the energy much more quickly, however, so its response is much quicker than the slow, deliberate mass-like inductor. The inductor current, then, lags the voltage, and the capacitor current leads the voltage.

Lighting ballasts use capacitance to store up enough charge to excite the neon or other gas to a state of sufficient energy to emit photons of light as electrons change orbitals.

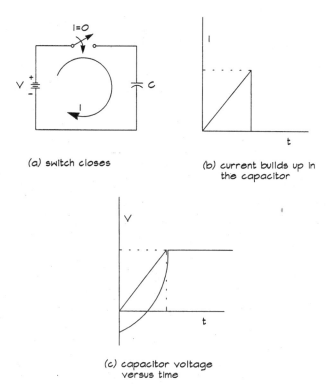

(a) switch closes

(b) current builds up in the capacitor

(c) capacitor voltage versus time

Figure 3.3 Characteristics of a capacitance circuit.

Impedance in Circuits

The next step is to create a physical model for a sinusoid to imitate the subtle but important aspects of wave propagation. Figure 3.4b shows a single pattern, slightly displaced from the origin. This kind of shift is caused by the presence of nonlinear elements in a circuit, such as inductance and capacitance. The shift is called the phase angle and is represented in Fig. 3.4a by a line or vector rotating at an angular frequency related to the period of the waveform by the function:

$$T = 2 \times \frac{\pi}{\omega} \tag{3.1}$$

where T = period
 ω = angular frequency

Notice in the illustration that the angle of displacement of the vector from the axis is equal to the displacement of the sine wave from the origin on the time axis (recall that a single wavelength is 360°).

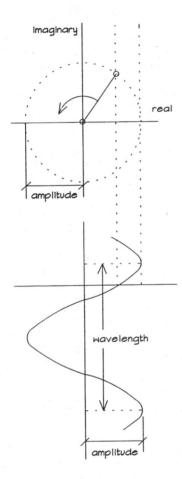

Figure 3.4 Projection of a rotating crank is a sinusoid.

The length of the vector is equal to the amplitude of the wave. The coordinate system for this vector is the horizontal real plane and the vertical imaginary plane. The axes are so named to indicate that the quantity symbolized by a single vector represents a sinusoidal wave of the given phase angle and amplitude.

Reactance on the Imaginary Plane

This new mathematical representation can now be used to denote the exact relationship between voltage and current in inductors and capacitors that are energized by an alternating voltage source. Notice in Fig. 3.5a that the voltage source is no longer a dc battery, but an alternating voltage in sinusoidal form. Instead of drawing separate longitudi-

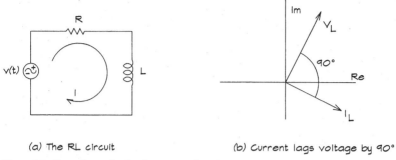

(a) The RL circuit (b) Current lags voltage by 90°

Figure 3.5 Impedance in the frequency domain.

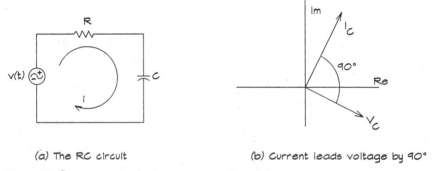

(a) The RC circuit (b) Current leads voltage by 90°

Figure 3.6 Capacitance in the frequency domain.

nal waves for current and voltage in each circuit element, these waves can be represented by vectors indicating amplitude and phase angle. Here the voltage through the resistor is shown on the real axis because it is in phase with the voltage source. However, the voltage through the inductor is lagging the voltage by 90°, as expected.

A similar analysis of a capacitance circuit excited by an alternating voltage source is shown in Fig. 3.6. Notice that the voltage is now lagging the voltage source, which is on the real plane. (The voltage in an inductance leads the source.) The vector representation of the RC circuit shows the current leading the voltage in the capacitance by 90°, as rationalized earlier.

Real and Reactive Power

The general case of power in ac circuits can now be presented with authenticity. Consider building service entrance power characteristics. The building will have many inductive loads (predominantly motors)

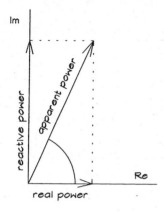

Figure 3.7 The power triangle.

and many loads that are mostly capacitance (e.g., fluorescent lamp ballasts). When all of these loads are combined, the result is a complex sum of the real and reactive power, as indicated in Fig. 3.7. The power triangle is the representation of three vectors in the real-imaginary coordinate plane: real power + reactive power = apparent power.

The apparent power is what you measure with meters at the service entrance and is what the electric bill is based on. The real power is what actually does the useful work in the system, turning motors, charging lights, and heating resistors. The reactive power, as the imaginary part of the apparent power, is proportional to the amount of energy that is given to the load by the ac source each cycle. This reactive power is positive for inductive loads and negative for capacitive loads.

More practically, the apparent power is what determines the actual operating level of motors, transistors, and all other equipment in the building. Losses in conductors are proportional to the square of the current, and losses in magnetic materials are proportional to the square of the voltage. For a steady-state power supply, apparent power limits imply both current and voltage limits. The result is that in a motor the actual work is done by the real power, but the apparent power is what determines the operating limits. This means a motor may be drawing 100 percent of power but only producing 90 percent as much work, or worse, if the service power factor is low. From another perspective, a motor required by its circuit to operate at 100 percent capacity will draw 100 percent of power and will be exceeding its operating limits while doing so.

The bottom line is that for systems with a large power factor (which is the angle between the apparent and real power vectors) equipment can become overloaded even when the load is within its operating parameters. Also, the reactive power is what is delivered from the power plant and billed for, even though it is not always used.

The Frequency Domain

The phasor representation of voltage and current in the real and imaginary planes is independent of frequency. The amplitude and phase angle (e.g., the starting position of the sinusoidal waveform) are indicated, but not the frequency. This would be indicated by the rate at which the phasor rotates about the origin, or zero point, of the coordinate system. This is usually not a problem because the frequency of the current and voltage are the same in a circuit excited from a common source.

The presence of harmonics in a system means there are voltages and currents in the circuit that are not at the base, or fundamental frequency. Usually they are a multiple of the fundamental. The third harmonic has 3 times the fundamental frequency, for example. The effect of this odd waveform on the circuit takes many forms. Figure 3.8 will help explain one ramification. It illustrates the energy flowing into a resistor as the result of excitation by a single-phase sinusoidal voltage. The average power is as indicated by the dashed line.

Imagine a second waveform overlaid on this pattern with 3 times the frequency, as indicative of the noise imposed by a third harmonic distortion in the circuit. This will add 3 times the power peaks to the plot and, depending on the amplitude, will increase the average power in the resistor. Thus there will be more energy lost from the resistor in the form of heat, and the general response of the rest of the circuit to the resistor may also be adversely affected.

Figure 3.9 shows the effect of harmonics on an inductor. As indicated in previous analyses, power is stored in an inductor in response to excitation by an alternating voltage source. This is a magnetic energy

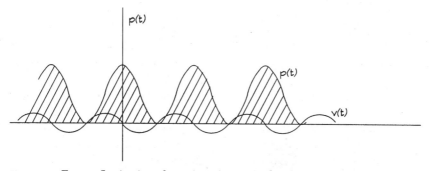

Figure 3.8 Energy flowing into the resistor (power is always positive).

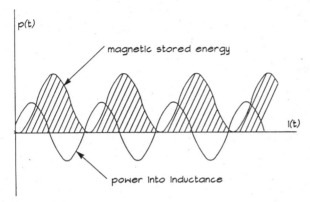

Figure 3.9 Stored energy and power in an inductor.

and is the same influence that causes the iron core of a motor to behave like a magnetic.

Imagine, again, a second waveform indicating the presence of a third harmonic in the circuit. There will be nine more peaks, spaced evenly on the graph. They will not only increase the average energy stored in the inductor but will cause higher peaks and dips of the overall power curve. This means more energy is retained in the circuit, doing no useful work, but only heating up the device and wasting energy in the form of thermal response. In practical terms, more energy is lost in the device with each repetition of the electromagnetic cycle.

In the special case of a motor, the difference between the peaks and the dips will be less. The torque that the magnetized stator can impart to the rotor is diminished proportionately, and the motor does less work. Also, because the dip is not quite to zero, the possibility exists that the corresponding dip on the rotor may be higher and there will actually be a reverse torque that is imparted to the device. Energy will have to be spent overcoming this influence, thus reducing the overall efficiency of the device.

These comments are made for a simplistic, single-phased system. A three-phase device is much more complex and designed to mitigate some of the adverse effects of harmonics. However, the effects cannot be completely avoided by design means, and the motor exhibits thermal and efficiency losses anyway.

Capacitors are affected in a similar manner by alternating current, storing a greater amount of energy than theory predicates. In motors the combination of the two effects are a portion of the inrush current that is needed to not only begin rotating the device against a load but also to fully energize the operating aspects of the electromagnetic device.

Power Factor Correction

As the previous discussion on motors indicates, one advantage of being conversant in the theoretical aspect of electricity is being able to comprehend some of the new, complex aspects of the phenomena. As another example, consider the means by which low power factors can be corrected by the addition of a capacitance to the circuit.

Large commercial and industrial facilities often have a large lagging power factor that can be corrected by the addition of a capacitance in parallel with the building load. The same can also be done for large motors, such as pump or electric chiller motors, to reduce individual power factors and that of the entire service as well.

The actual operation of the capacitance is simple. As described earlier, when first energized, the capacitor draws current from the source until fully charged. Then, as the voltage drops, it discharges to the circuit, then recharges, and so forth. With each subsequent cycle, though, the building service is charged from the capacitance and not from the power company service. The energy is stored by the capacitor and kept locally for use as required, like a flywheel storing energy as a vehicle goes downhill, then discharging it to the transmission to help go up the next hill.

Safety

The effect of alternating voltage excitation on even these simple circuit elements is a complex phenomenon. Circuits are characterized by their simple influence on the charging waveforms and also by their ability to store and discharge electrical energy. This, in turn, is a function of time. A disabled circuit may still retain charge that can be dangerous to electricians working on the equipment.

Chapter 4

Power Circuiting

This chapter looks at how power is delivered to a building and then distributed to all the points of use. The first section looks at the service entrance and main feeder to the building. The next section locates the outlets, junction boxes, and disconnects at the points of use. The last section discusses the circuiting and panel locations.

Temporary Service

All construction sites require temporary service, and the utility will usually provide 120/240-V, three-wire, single-phase service. The minimum wire size for the installation should be #8 AWG copper wire, or higher if the load is greater. Figure 4.1 is an illustration of a typical temporary service. The treated pole should be in the ground at least 3 ft deep, and at least three braces (not shown) should be used to support the meter pole. The contractor and the owner are responsible for the safety of this installation. It is best to build a fence to prevent any activity within 6 ft horizontally of the service lines, in addition to proven safeguards such as ground fault protection. Only qualified electrical personnel should be allowed access to this enclosure.

Service Installation

An important part of the electrical distribution system is the wiring from the utility's supply point to the beginning of the customer's utilization wiring. This section first reviews the requirements for a typical overhead service for small to medium-sized loads. The maximum overhead service is as follows:

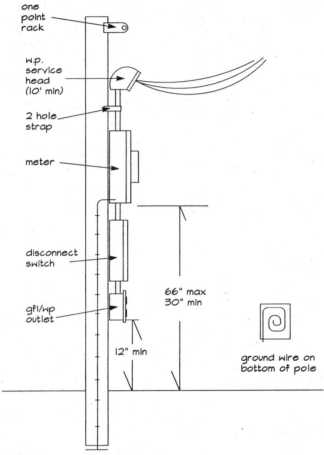

Figure 4.1 Portable meter loop for temporary overhead construction service.

Voltage (V)	Size of service (A)
120/240, single-phase, three-wire	800
120/240, three-phase, four-wire, delta	1600
120/208, three-phase, four-wire, wye	1600
240/480, single-phase	200
480, three-wire, delta	800

All other services at either a higher voltage or amperage must be routed underground to the point of use. Table 4.1 gives a range of service capacities for small commercial installations, and their power requirements.

TABLE 4.1 Nominal Service Size in Amperes*

Facility	1000 (ft²)	2000 (ft²)	5000 (ft²)	10,000 (ft²)	Remarks
Single-phase 120/240-V, three-wire					
Residence	100	100	200		Min. 100
Store	100	150			
School	100	100			
Church	100	150	200		
Three-phase 120/208-V, four-wire					
Apartment house			150	150	
Hospital			200	400	
Office			400	600	
Store		100	400	600	
School		400	150	200	

*Nominal service sizes are 100, 150, 200, 400, 800, 1200, 1600, and 2000 A.
SOURCE: Reprinted with permission of NFPA 70-1993, National Electric Code.

Another helpful way to size service for a building before the plans are complete is to refer to Table 4.2, which lists some maximum values for power use in different facilities and types of spaces. Notice the final column and the expected load growth that should be added to the building service capacity. The values in the table can also be considered a design maximum for each space type. If the final design is greater in any area, there must be some unusual reason, such as high computer density or other special requirements.

Typically, the utility provides only one service point and one voltage to a single building (i.e., a continuous structure or a group of buildings joined into a single entity) or single tract of land. Additional services are provided at the customer's expense and at the discretion of the utility. One common exception to this rule applies to multiple occupancies such as a small strip shopping center. (An acceptable meter bank enclosure is shown in plan and section view in Fig. 4.2.) They are still limited to a single service point at one voltage. However, if there is a lease space with a high load at a different voltage such as a grocery store with 1000 A of refrigerating equipment at 277/480 V, a second service is usually provided.

When there are two separate services to a building, at different voltages, it is very important that they be completely separated. The higher-voltage service, in particular, should be in a 1-h-rated enclosure, and there should be no accesses, such as windows or access panels, between the two rooms at different voltages. In addition no conduits of different voltage should pass through either electric space. These

TABLE 4.2 Electric Load Estimating

Type of occupancy	Voltamperes per square foot				10-year % load growth
	Lighting demand	Misc. power	Electric ac	Nonelectric ac	
Auditorium					
General	2–3	0	15–22	6–10	20–40
Stage	20–40	.5			
Art gallery	4.5–6	1	5–7	2–3.2	20–40
Bank	3.5–4.5	2	6–10	2.5–3.2	30–50
Cafeteria	5–6	1	5–7	2–3.2	20–40
Church	2–4	.5	6–10	2.5–3.2	10–30
Computer area	4–6	2	15–30	4–12	50–200
Department store:					
Basement	5–6	1.5			
Main floor	4–5	1	6–10	2.5–3.2	50–100
Upper floor	3–4	.5			
Dwelling:					
0 to 3000 ft^2	3	.5			
3000 to 120,000 ft^2	2	.25			50–100
120,000+	1	.15			
Garage (commercial)	0.5	.15			10–30
Hospital	2–3	1	5–7	2–3.2	40–80
Hotel:					
Lobby	6–8	.5	8–10	4–5.5	
Rooms	1.5–2.5	.5	5–7	2–3.2	30–60
Industrial loft building	2–3	2			
Laboratories	5–7	5–20	6–10	2.5–4.5	100–200
Library	4–7	1	5.5–7.5	2.2–3.5	100–300
Medical center	4–6	2	4–7	2–3.2	30–40
Motel	1.5–2	.5			50–80
Office building	5–7	2	5.5–7.5	2.2–3.5	30–60
Restaurant	2–3	.25	6–10	2.5–4.5	20–40
Schools	3–5	2	5.5–7.5	2.2–3.5	50–80
Shops:					
Barber and beauty	5–7	1	6–11	2.4–5.6	
Dress	4–7	.5			
Drug	5	.5			40–80
Five and ten	4	1	5–9	2.4	
Hat, shoe, specialty	4	.5			
Warehouse (storage)	.25–1	.25			10–30
In the above except single dwellings:					
Corridors, closets	.5				
Storage spaces	.25				

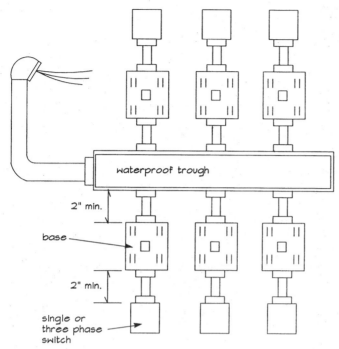

Figure 4.2 Multiple meter installations.

rules are important to keep the services isolated, thus minimizing the possibility of cross-connections when repairs or additional loads are added to the service.

It is expected that single-phase 120/240-V, three-wire service will be the customer's standard for lighting, general appliances, and motors. Larger commercial installations with a significant load, 100 A or more, usually qualify for three-phase 120/240-V, four-wire service. A motor load or straight power load of 100 A or more is also needed for a three-phase 480-V four-wire service.

The service entrance conductors should be color coded according to a standard established by the utility company. First the phases must be established by a specific criteria, such as

> Phases shall be arranged in the order A, B, C from front to back, top to bottom, or left to right as viewed from the front of the meter, panel or other service equipment device.

Then, following this standard, the color coding standard can be set up for all conductors at all terminals for wire sizes #10 AGW and larger:

Conductor	Phase A	Phase B	Phase C	Neutral
120/240-V, single-phase, three-wire	Red	Black		White
120/208-V, three-phase, four-wire	Red	Black	Blue	White
120/240-V, three-phase, four-wire, delta	Red	Orange (high leg)	Black	White
Metering equipment	Red	Black	Orange (high leg)	White
Green is used for grounding conductors, only				

The utility can provide power within 5 percent of the nominal voltage most of the time. Most equipment can operate without any problems in this voltage range. Some equipment, such as computer or controls networks, require a higher quality of power. It is usually the customer's responsibility to provide uninterruptible power supplies for this purpose. It is also the customer's responsibility to install devices to protect motors from single-phasing, due to a fault in either the utility's service or the facility's supply lines.

Service drop conductors

The service drop wires are what connect the utility's supply source with the customer's service entrance conductors. There are strict NEC guidelines for the allowable height of these conductors above the finished or final grade. These are provided in Table 4.3.

Another parameter is that an 8-ft clearance is needed over the highest point of the roof; there are a few exceptions that allow lower heights. It is always best to install the wires as high as possible. This will not only allow for some thermal expansion of the wire during the hottest months but will also provide the safest possible installation from interference from vehicles of all sorts.

Service entrance

A service entrance for above-roof installations is shown in Fig. 4.3b (Fig. 4.3a shows an installation under the eaves of a building). The overhead conductors are anchored to the building by a one-, two-, three-, or four-point rack. Each assembly is described as follows:

One-point rack, for three- and four-wire service of 400 A or less, except 480 V. The rack must be installed below and within 10 in of the weatherhead.

Three-point rack, for all three-wire services rated above 400 A and

TABLE 4.3 Overhead Service Clearances from Ground

Height	Application
Above roofs*	Conductors shall have a vertical clearance of not less than 8 ft (2.44 m) above the roof surface. This clearance shall be maintained for a distance not less than 3 ft (914 mm) in all directions from the edge of the roof.
10 ft (3.05 m)	At the service entrance, at the drip loop, or above areas accessible only to pedestrians. Only for service entrance feeder cabled together with a grounded bare messenger and limited to 150 V to ground.
12 ft (3.66 m)	Over residential property and driveways and those commercial areas not subject to truck traffic. Voltage is limited to 300 V to ground.
15 ft (4.57 m)	For those areas listed in the 12-ft (3.66 m) classification where the voltage exceeds 300 V to ground.
18 ft (5.49 m)	Over public streets, alleys, roads, parking areas subject to truck traffic, driveways on other than residential property, and other land traveled by vehicles such as cultivated, gracing, forest, and orchard.

*There are several exceptions to the first, "Above Roofs," category:
 1. If the area above the roof is subject to pedestrian or vehicular traffic, the clearances in the rest of the table shall apply.
 2. Where the voltage does not exceed 300 V and the roof has a slope not less than 4 in in 12, a reduction in clearance to 3 ft (914 mm) is permitted.
 3. Where voltage between conductors does not exceed 300 V, a reduction in clearance above the overhanging portion of the roof to not less than 18 in (457 mm) is permitted if (*a*) not more than 6 ft (1.83 mm) of service-drop conductors, 4 ft (1.22 m) horizontally, pass above the roof overhang and (*b*) they are terminated at a through-the-roof raceway or approved support.
 4. The requirement for maintaining the vertical clearance 3 ft (914 mm) from the edge of the roof shall not apply to the final conductor span where the service drop is attached to the side of a building.
Source: Reprinted with permission of NFPA 70-1993, National Electric Code.

all 480 V three-wire services regardless of ampacity. The highest rack, when installed vertically, must be installed no more than 10 in below the weatherhead. The centerpoint of the rack must be directly below the weatherhead(s). The racks shall be 10 to 12 in apart, on center.

Four-point rack, for all four-wire services rated over 400 A and following the same orientations as for a three-point rack.

The dead end attachment, or rack, is usually installed at the customer's expense and at a location agreed upon by the utility. The attachment must be to the structural frame of the building. If the building is wood frame, the point of attachment must be a header, of at least 2- × 6-in construction. All fittings and attachments must be weatherproof and galvanized. Any welds made must be repaired to the original condition.

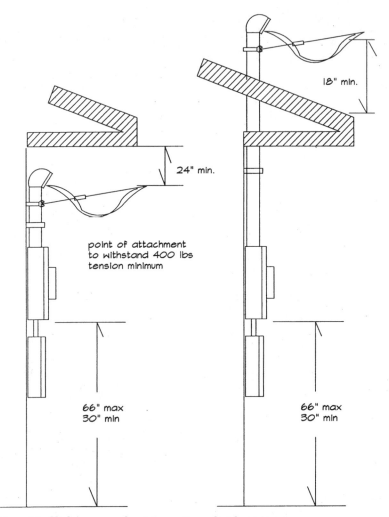

Figure 4.3 Undereaves and service mast overhead services.

The next element of the electrical service is the service conduit. The conductors are usually service entrance grade and are installed in rigid metal conduit from the weatherhead to a junction box or trough. The service conduit must be securely attached to the building or to kindorf or unistrut racks if the conduit must stand off the building. All straps need to be secured to structural members of the building frame. The box or trough is sized according to the number of conduits, as in Table 4.4. It should be installed at least 12 in and no more than 42 in above final grade. Underground service is terminated in a trough by similar means.

TABLE 4.4 Junction Box and Trough Specifications

Conduits to junction box from transformer (in)	Minimum junction box size (in)	Minimum trough size (in)
1–2	18 × 18 × 8	6 × 6 × 48
2–2	18 × 18 × 8	8 × 8 × 48
1–3	18 × 18 × 8	8 × 18 × 48
2–3	24 × 18 × 10	8 × 8 × 72
1–4	24 × 24 × 10	8 × 8 × 96
2–4	30 × 30 × 12	8 × 8 × 96
3–3	30 × 30 × 12	10 × 10 × 96
4–4	42 × 42 × 14	12 × 12 × 96
6–4	48 × 48 × 14	12 × 12 × 96

Notes:
1. Service junction boxes installed ahead of metering equipment or ahead of the main service disconnect are sized by the number and size of conduits from the transformer or service. Troughs are sized by the size of the junction box.
2. The minimum feeder wire installed from the trough into the junction box must be sized to meet a minimum load of 15 W/ft^2 for multifuel locations or 20 W/ft^2 for all electric buildings. This applies to that portion of the building that will be served by the trough.
3. Terminal blocks are required in all new troughs and junction boxes for both commercial and residential services. Split bolt connectors are not allowed in new troughs, existing troughs, or junction boxes that have terminal blocks. The terminal blocks shall be sized to meet the existing and anticipated future loads and shall meet all the applicable ANSI/EIA standards for electrical connectors.

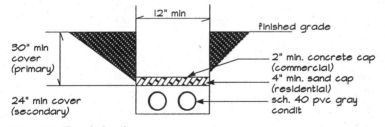

Figure 4.4 Trench detail.

Underground service conduit should be rigid metal conduit, galvanized and suitable for underground service. The conduit should be buried at least 24 in deep, measured from the top of the conduit to the finished grade. This minimum distance must be maintained in all circumstances, even if there is a poured concrete pad above. Refer to Fig. 4.4 for details on underground trenching requirements.

Terminal blocks are installed in the trough or junction box. The utility uses them to attach the service lateral conductors to the customer's service entrance conductors. The terminal blocks should be sized for the largest expected service to the building in the future.

The main building disconnect switch is next in line. It should be sized for the entire future expected service and fused to the current service level. In buildings with many meters, there should be one fused disconnect ahead of each of them. All main service disconnects must be in lockable enclosures.

The building electric meter and current transformer, if required, are installed near the service entrance and disconnect. The power then goes to the main distribution panel of the building, from which it is delivered throughout the facility.

Points of Use

One of the first things that the electrical designer should require of the client is the location of all major items of equipment. Everything from computers to elevator motors, copy machines to sterilizers, microwaves to steamers should be included in this inventory. If the owner knows the power requirements, that will help too. If it is existing equipment to be relocated, the nameplate data can be used for the voltage and full-load amps. Otherwise a standard table such as Table 4.5 can be used for this information.

TABLE 4.5 Power Consumed by Appliances (Average)

Appliance	Watts	Appliance	Watts
Air conditioner, room type	800–1500	Mixer, food	150
		Motor, per horsepower	1000
Blanket, electric	175	Oven, built-in	4000
Broiler, rotisserie	1400	Radio	75
Clock, electric	2	Razor	10
Coffee maker	600	Range (all burners, oven on)	8–16,000
Dishwasher	1800		
Dryer, clothes	4500	Range, separate	5000
Fan, portable	175	Refrigerator	250
Freezer	400	Roaster	1380
Fryer, deep-fat	1320	Sewing machine	75
Frying pan	1000	Stereo	300
Garbage disposer	900	Sun lamp (ultraviolet)	275
Heater, portable	1200	Television	250
Heater, permanent wall type	1600	Toaster	1100
		Vacuum cleaner	400
Heat lamp (infrared)	250	Waffle iron	800
Heating pad	75	Washer, automatic	700
Hot plate (per burner)	825	Washer, electric, manual	400
Iron, hand	1000		
Iron, motorized	1650	Water heater, standard 80 gal	4500

All of this information should then be put on the building plans, typically with the use of keyed notes. If there is a great deal of equipment, such as for a restaurant or manufacturing facility, a full schedule is a more effective way to present the information. This information should also include the type of outlet or if a junction box is needed for a hard-wired connection. Either way, the more information that can be included on the plans the better.

One benefit of assimilating information in this way is if a wall or cabinet or entire space is moved late in the design process, it will be easy to identify all of the power requirements that go with the move. Another benefit is that the very act of requesting this information from the client, and then the process of organizing and sorting it on the plans, will cause both the client and the designer to address all of the pertinent issues. For example, it may be discovered that an extra shelf or counter will be needed for a piece of equipment or that it will need a dedicated outlet. Other items such as a fax machine or copier may require a computer network connection or a phone line. Physically locating each piece of equipment makes it possible to provide all of its power and signal needs early in the design process.

Another important use for this information is in the design of the hvac system. The mechanical designer needs to know equipment locations and power use so that the cooling load for the space can be increased accordingly. Otherwise, the room will not be conditioned properly, and the occupants may suffer and the equipment may be damaged as well by operating at elevated temperatures.

Once the major items of equipment have been located, those that require a dedicated circuit can be circuited immediately. Any circuit with a freezer or refrigerator should be on a dedicated circuit, regardless of size or power use. The reason for this is that when the compressor cycles on, there is a momentary drop in the voltage of the entire circuit. If computers or similar equipment are on the same circuit, they could have problems such as lost data or a lost network connection. This can happen with almost any circuit that includes a compressor or motor and that cycles on and off frequently. The motor draws extra power from the circuit when starting. If the motor is very large, it could even affect the entire panel, which is why such equipment is typically circuited on a completely separate equipment branch.

The next task is to locate all the duplex outlets. It helps to know what the space use will be. Offices should have at least one outlet per wall, spaced so that any possible furniture arrangement can be accommodated. Centering the outlet on the wall is sufficient in many cases, except where there is a door or door swing or any other type of dead space such as a glass wall. It is important to get input from the client on office outlets and to identify those most likely to be nearest desks. These should be provided with a telephone jack nearby, plus a computer jack for either current or future use. If the furnishings are to be modular furniture with integral wiring, often all that is required is a junction box at the proper location for hard wiring to a dedicated circuit.

Large open spaces that will be partitioned into offices are a little more challenging. Most perimeter offices can be supplied via wall outlets. Inner offices can have floor outlets, which are fairly expensive and should be used only if access is not possible by any other means. Power poles are a better alternative because they are inexpensive and can be moved to fit the office furniture configuration. They can also be provided with telephone and computer outlets. Finally, if the area is very computer intensive and with many work spaces, such as a news room or telephone marketing office, it may be more economical to install a raised floor and to serve power and communications from floor outlets.

Some other important things to keep in mind are

- All outlets should be installed at 18 in above finished floor, for full accessibility.
- Any outlet near a sink or potential water source should be ground-fault protected.
- Safety outlets should be specified in public areas, especially where small children may be.
- The owner should be asked if any outlets for seasonal use are required, such as holiday lights or for special sales signs or exterior illumination. This is especially important in retail applications.

Once all of the outlet locations have been established, and approved by the owner, it is helpful to check the results before circuiting them. Table 4.6 provides a conservative goal for average receptacle power densities in different building types.

It is always good practice to have more than one circuit supplying each room, for backup in the event power to one of the circuits is lost. It

TABLE 4.6 Average Power Receptacle Power Densities

Building type	Average (W/ft^2)
Assembly	0.25
Office	0.75
Retail	0.25
Warehouse	.010
School	0.50
Hotel or motel	0.25
Restaurant	.010
Health	1.00
Multi-family	0.75

is conservative to assume 250 to 400 W per outlet, with a maximum of six to eight outlets per 120-V 20-A circuit. If it is a computer-intensive operation, the more conservative figures should be used as a maximum. Note that as computers become more powerful, their power requirements are increasing, not to mention larger monitors and the addition of more peripheral devices such as modems, digitizers, and local printers.

The last equipment that is usually circuited on a job is the hvac equipment—cooling fans, air handlers, chillers, and exhaust fans. Each item should have a disconnect switch near the equipment (never mounted to it, because that makes replacement and repairs difficult), within sight, and as close as possible.

In larger facilities it is helpful to have separate panels for the following feeders so that power used by each can be monitored independently:

- Lighting and duplex outlets
- The hvac systems and associated equipment
- Large motors and special equipment in kitchens, computer rooms, and manufacturing areas

It is best to keep other loads on each feeder to a maximum of 10 percent. This should be a requirement for service with a connected load over 250 kVA. Even if the branches are not initially metered, the ability to do so can be important in retrofit projects. In addition, a current loop can be used to monitor usage via an energy management system.

Table 4.7 provides a helpful guideline for the locations in which different types of conduit can be used, as a minimum standard of compliance.

One-Line Riser Diagrams

Somewhere in the process of selecting equipment, panels, and major service points, it can get quite complicated. Small projects are sometimes more confusing than large ones, especially if there will be a new or upgraded service required at the installation. A very good way to sort it all out is to construct a one-line diagram. This is a simple illustration showing all of the major components of the power distribution system:

- Service entrance configuration, with ct-can, meter, and weatherhead
- Main distribution panel with feeders to all dedicated (nonpanel) loads
- Transformers
- Subpanels and their respective feeders

The idea is to show the voltage and rated amps of all of these major

TABLE 4.7 Uses for Steel Raceways

NEC article	Raceway type	1*	2	3	4	5	6	7	8	9	10	11	12	13	14	15
346	Rigid steel conduit, zinc-coated	P†	P	P	E	P	P	P	P	P	P	P	P	P	P	P
346	Rigid steel conduit, enamelled	P	X	X	X	E	X	P	P	X	P	P	P	P	P	X
348	Electrical metallic tubing	P	P	P	E	P	P	P	E	P	E	P	X	P	P	P
350	Flexible steel conduit	P	E	E	E	E	E	P	E	X	E	P	X	P	P	E
352	Surface steel raceways	P	X	X	X	X	X	X	X	X	X	P	X	P	E	X
354	Underfloor steel raceways	P	—	—	—	P	—	P	X	X	X	—	—	X	P	—
356	Cellular steel raceways	P	—	—	—	P	—	P	X	X	X	—	—	—	P	—
362	Steel wireways	P	E	X	X	X	P	P	X	X	X	P	X	P	X	X
364	Steel busways	P	E	X	X	X	E	P	X	X	X	P	X	P	X	E
351	Liquid tight flexible conduit	Limited to use for connection of portable equipment or for motors where flexibility is needed.															
374	Auxiliary steel gutters	Limited to use as supplemental wiring space for services															

*Column key:
1. Inside building
2. Outside building
3. Underground
4. Cinder fill
5. Embedded in concrete
6. Wet locations
7. Dry locations
8. Corrosive locations
9. Severe corrosive locations
10. Hazardous locations
11. Mechanical injury
12. Severe mechanical injury
13. Exposed work
14. Concealed work
15. Services outside

†Permitted locations key:
P = Permitted
E = Exception
X = Not permitted
— = Conditions do not apply

components, with the wire sized to each device. A single illustration should be sufficient to show everything so that the permitting authorities can see that the service is sized properly.

One of the most important features of the one-line diagram is to size all feeders and conduits to the equipment shown. This makes the system much easier to review than if all the sizes are scattered through the drawings at random locations. In fact, it is better to provide sizing only in the one location, on the riser diagram, and nowhere else. That way if an element must be resized late in the design process, there is only one place where the value must be changed on the drawings. In the rush of last minute changes, this reduces the

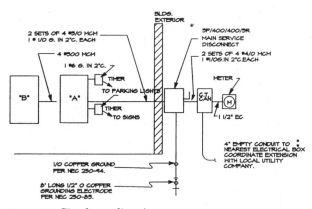

Figure 4.5 Simple one-line riser.

time to make the changes and also minimizes the potential for having conflicting numbers on the drawings.

It is always appropriate to have a one-line diagram, regardless of the size of the project. Figure 4.5 illustrates a simple facility, and Fig. 4.6 shows a more complex one with emergency power requirements. The one-line diagram should provide a snapshot of the whole project, with feeders shown to all panels as well as to all large amperage equipment such as elevator motors, chillers, and air handlers. There are several things that should be clear for each feeder:

- Wire size, number of wires, grounding wire, conduit size
- Disconnecting means—a disconnect switch, circuit breaker, or a motor control center element—plus, in all cases, the rating and the fuse rating (as applicable)
- The rated amps of each point of use (the service capacity in the case of panels, and not the actual design load)

Figure 4.7 lists commonly used symbols on more complex one-line electrical diagrams, such as the natural gas engine generator installation used for peak load reduction illustrated in Fig. 4.8.

Schedules

The electrical equipment schedules should also be in context with the riser diagram. These include panel schedules for the main distribution panel (MDP), the motor control center (MCC), and all power and lighting panels (Fig. 4.9). Ideally all of these schedules should be located on the same sheet as the one-line riser, or at a minimum the

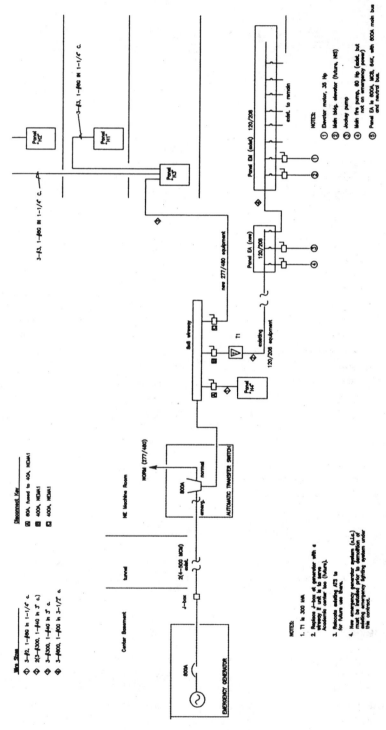

Figure 4.6 Complex one-line riser diagram.

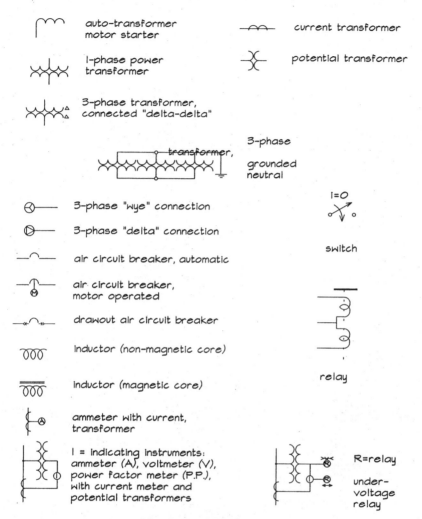

Figure 4.7 Symbols for one-line electrical diagrams.

MDP and MCC should be. This is very helpful in the development of the design and in making any late changes to it.

It is important that the actual loads be totaled for each type of schedule, using appropriate diversity factors. In the case of power it is not unreasonable to assume close to 100 percent loading of the system to account for future technology use at each workstation. Foresighted electrical designers who oversized capacity in new buildings 10 or 20 years ago have done the owners a great service. The power requirements of all their tenants can still be met without costly upgrades, which would have been done at a time late in the building's useful lifetime. Such upgrades are difficult to justify in terms of economic

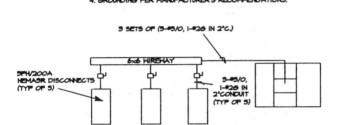

NOTES

1. ALL DISCONNECTS IN LINE OF SIGHT, 30'
2. FINAL CONNECTION TO UNITS VIA MIN 4'0" FLEXIBLE METAL CONDUIT.
3. WIREWAY, NEMA 3R RATED.
4. GROUNDING PER MANUFACTURER'S RECOMMENDATIONS.

Figure 4.8 Emergency genset on-line diagram.

PANEL SCHEDULE B

120/240 VOLTS — SERVES EQUIPMENT — 3 WIRE, 1 PHASE, SURFACE MOUNTED — 225 MLO, CCT SPACES −42 — 225A BUSS, NEMA 1 ENCLOSURE

CCT NO	SERVICE	VA AØ	VA BØ	VA CØ	BRKR AMP	POLE	WIRE SIZE FOR THHN COND	SHOWN IS COPPER WIRE	COND	BRKR AMP	POLE	VA AØ	VA BØ	VA CØ	SERVICE	CCT NO
1	SOUP WARMERS	600			20	1	12	1/2	12	1/2	20	1	100			WATER HTR CONTROLS	2
3	DRINK DISPENSER		2760		30		10							360		RECEPTACLES	4
5	CASH REGISTER			48	20		12	1/2							618	AMP & TIME CLOC	6
7	FUTURE CASH REG.	48									15	2	800			ICE MKR/REM COND.	8
9	CASH REGISTER			48										600			10
11	DELI CASE COND.		1850								20	1			1800	SOUP WARMER	12
13	DELI CASE EVAP.	1800												690		BEVERAGE SYSTEM	14
15	ESPRESSO COFFEE		1670		30									1800		FREEZER	16
17	COFFEE GRINDER			288	20										540	RECEPTACLES	18
19	SLICER & SCALE	700			20								700			REM PRINT&MULTIP.	20
21	CHEESE MELTER		600		20									1200		PROOFER	22
23	FINISH TABLE			660	20				8	3/4	50	3			4380	CONVECTION AVON	24
25	DRIVE THRU DISPENSER	240			20		8	3/4	12	1/2			4380				26
27	DRIVE-THRU WARMER		1800		20		10	1/2						4380			28
29	INTERCOM MASTER			110	20		12	1/2			20	1		984		MIXER	30
31	DRIVE THRU REGISTER	48			20								1500			DRIVE-THRU MENU BOARD	32
33	PIZZA TABLE		660		20									1404		WALK-IN COOLER	34
35	PIZZA OVEN			2808											120	WALK-IN COOLER LIGHT	36
37		2808														SPARE	38
39	MEAT & CHEESE TAB.		360														40
41	SPARE																42
	TOTAL	AØ 14414	BØ 17642	CØ 14206	PANEL TOTAL	KVA 46.3				AMPS 128.5		FEEDER REF. RISER DIAGRAM					

Figure 4.9 Panelboard schedule.

payback and are very expensive. Facilities with extra capacity already are the most sought after and the most fully occupied. (Highly computerized businesses that require such service are also those most able to pay the highest lease costs for office space.)

All of the panels should be scheduled with equal diligence. Each panel should have a minimum of 10 percent spares and 30 percent spaces for future expansion. The panel rating and feeder should be sized for this anticipated extra capacity. The formal panelboard schedule should be filled out completely, with all known loads identified clearly. It is helpful to have the total load in amps provided for each home run, plus the wire size calculated for a maximum 2 percent voltage drop.

There are many benefits to providing so much information on the plans. First, the data must be calculated anyway to size wiring and organize the home runs. Second, if the design is complete, the electrical contractor will be less inclined to change anything or to increase the bid price to accommodate this contingency in the construction costs. Third, it makes the designer double-check everything before the bid documents are issued. Error-free bid documents are invaluable, and the little extra effort that it takes to produce them is always less than the trouble and anxiety needed to issue addenda. Finally, it is a very good service to the contractor and of great benefit to field personnel who actually perform the work. Thorough panel schedule designs allow them more opportunity to concentrate on balancing the phases, which is extremely important.

Phasor Representations

Using the vector mathematics developed in the previous chapter, the power system can be presented in a more comprehensive manner. The analysis to date is limited, in that the conceptualization is based on the simple linear conception of a three-phase source. The three phases are shown between the main conductors in Fig. 4.10.

The three voltages are shown in the time domain in Fig. 4.10a. This shows three sinusoidal waves that are 120° apart, which hint at a smooth energy transmission as a result of the multiple-phase system. Their phase relationship is shown in Fig. 4.10b, in the frequency domain.

The phase relationship of a delta-connected load is shown in Fig. 4.11a. The three phases connected by the alternating current symbol indicate the delta source, and the load is similarly connected. The phasor diagram is in Fig. 4.11b, indicating the phase relationship between the three phases of power supplied.

Notice that the phases are symbolically balanced, the amplitude of each phase being equal. The net result is a zero-voltage neutral, and no neutral current occurs. This is indicated in Fig. 4.11, which shows

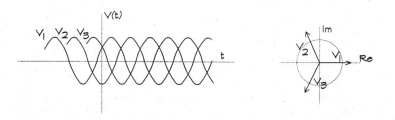

(a) the time domain   (b) the frequency domain

Figure 4.10 Voltages of a three-phase system.

62 Chapter Four

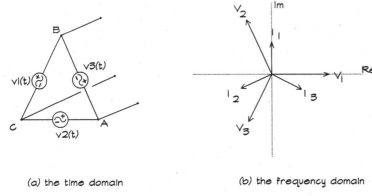

(a) the time domain

(b) the frequency domain

Figure 4.11 Three-phase relationships.

the line voltages and line currents in this balanced circuit. The phase current lags the phase voltage in each leg because this circuit has a 0.9 lagging power factor.

There are two things to consider now. First, unless it is a dedicated circuit for a motor, actual electrical circuits never have perfectly balanced phases. This means the voltage phasors will be of different lengths, and their respective phase currents will also be of different lengths. Summing these currents will not result in a zero vector, but a neutral current. Second, the relative impedance of each phase will be different. Thus, even if the phase currents were perfectly balanced, they would not sum to zero because they would not be 120° out of phase. These two circumstances mean that in virtually every circuit there will be a neutral current.

This shows how important it is to balance loads to the extent possible. It also supports the ideal of using separate panels for power, lighting, and equipment circuits that have predominantly resistance, capacitance, and inductance loads, respectively. If all loads are lagging on a panel, for example, balancing the current ratings on each phase results in an approximate balance of the loads because they are all displaced from the normal by roughly the same amount. If there are lagging and leading loads on the same panel, however, the actual currents will vary in different directions from the line current, so balancing the currents on each phase of the panel is virtually impossible.

Another advantage of separating the loads in this way is that capacitors can be used for power factor correction at the panel itself. Otherwise, if the correction is done at the service point, the entire facility will still experience the symptoms of the poor power factor.

Figure 4.12 provides some typical power symbols and Figs. 4.13 and 4.14 are two typical equipment schedules.

Power Circuiting

WIRING DEVICES

⊖_c	SIMPLEX RECEPTACLE (INDICATES CLOCK HANGER RECEPTACLE)
⊖	DUPLEX RECEPTACLE.
⊖	DUPLEX RECEPTACLE WITH ISOLATED GROUND
⊖_WP	DUPLEX RECEPTACLE WITH INTERNAL GROUND FAULT PROTECTION. (W.P. INDICATES WEATHERPROOF)
⊖	DUPLEX RECEPTACLE. MOUNT 48" A.F.F. OR 6" ABOVE COUNTER TOP WHERE COUNTER IS INDICATED.
⊕	DUPLEX RECEPTACLE ON EMERGENCY CIRCUIT
⊕	QUADRAPLEX RECEPTACLE.
⊙	FLUSH FLOOR DUPLEX RECEPTACLE.
⊙▼	FLUSH WITH FLOOR COMBINATION DUPLEX RECEPTACLE AND TELEPHONE RECEPTACLE.
⊙▼▼	FLUSH WITH FLOOR COMBINATION DUPLEX RECEPTACLE TELEPHONE AND DATA RECEPTACLES.
	SPECIAL PURPOSE RECEPTACLE. SIZE AND NEMA CONFIGURATION AS INDICATED ON DRAWING.

POWER

J	JUNCTION BOX WITH COVER PLATE
E	EQUIPMENT CONNECTION. (PROVIDE ALL BRANCH CIRCUITRY REQUIRED TO CONNECT TO EQUIPMENT)
(1/2)	MOTOR CONNECTION, HP INDICATED.
▬	PANELBOARD (208/120 VOLT).
C	ENCLOSED CIRCUIT BREAKER MOUNT 54" A.F.F.
☐ 3P/60/40/5R	FUSED DISCONNECT SWITCH. IN NEMA I ENCLOSURE U.O.N. 3P = NO. OF POLES, 60 = SWITCH RATING, 40 = FUSE RATING 5R = NEMA 5R
☐ 3P/60/5R	NON-FUSED DISCONNECT SWITCH. IN NEMA I ENCLOSURE U.O.N. 3P = NO. OF POLES, 60 = SWITCH RATING, 5R = NEMA 5R
⊠ S1	MAGNETIC MOTOR STARTER. NEMA I ENCLOSURE U.O.N. MINIMUM NEMA SIZE I.
⊠	COMBINATION MAGNETIC STARTER AND DISCONNECT.
S_M	MANUAL MOTOR STARTER WITH THERMAL OVERLOADS.
[C]	CONTACTOR, RATING AND POLES AS INDICATED.
[TC]	TIME CLOCK, RATING AS INDICATED.
[PC]	PHOTO-ELECTRIC CONTROL.
[•]	PUSHBUTTON

Figure 4.12 Power symbols.

DISTRIBUTION PANELBOARD SCHEDULE DP

BUS: 400 AMPS, 3 PHASE, 4 & GR. WIRE 277/480 VOLTS
U.L. SHORT CIRCUIT RATING: 30K R.M.S. SYM. AMPS.
MAINS: M.L.O. , 400 AMPS, 3 POLES
MOUNTING SURFACE , NEMA 1 ENCLOSURE

CKT. NO.	SERVES	BREAKERS			FEEDER	
		FRAME AMP	POLE	TRIP	NO. & SIZE WIRE	COND.
1	PANEL HK	400	3	250	4-#300, 1-#4 GR.	3"
2	TRANSFORMER TA	225	3	150	3-#8, 1-#10 GR.	2"
3	PANEL HB	225	3	100	4-#1, 1-#8 GR.	2"
4	PANEL HC	225	3	150	4-#3/0, 1-#6 GR.	2"
5	R/F UNIT	225	3	125	--	3"
6	CT SCAN UNIT	225	3	175	--	3"
7	ULTRASOUND UNIT	100	3	100	--	3"
8	SPARES	100	3	100	--	-
9-11	SPACES	100	3	-	--	-

Figure 4.13 Distribution panelboard schedule.

MAIN SWITCHBOARD SCHEDULE - MSB

LOCATION: MECH ROOM
BUS: 1000 AMPS 3 PHASE 4 & GR. WIRE 277/480 VOLTS
M.L.O. / 1000 AMPS - BRACED AT 42K RMS SYM AMPS.

CKT. NO.	SERVES	CKT. BREAKER			FEEDER	
		FRAME AMP	POLE	TRIP	NO. & SIZE CONDUCTOR	COND.
1	DIST. PANEL DP	400	3	400	4-#500, 1-#3 GR.	3-1/2"
2	CHILLER #1	400	3	300	3-#500, 1-#3 GR.	3"
3	CHILLER #2	400	3	200	3-#500, 1-#3 GR.	3"
4	ATS - EMERGENCY	225	3	225	4-#3/0, 1-#6 GR.	3"
5	MOTOR STARTER PANEL	100	3	-	3-#1, 1-#8 GR.	4"
6	SPARE	225	3	200	--	2"

Figure 4.14 Main switchboard schedule.

Chapter

5

Signals Circuiting

A solid knowledge of the signal requirements for a facility is a valuable asset to an electrical designer. Individually each signal and communication system is not complex because they all follow some relatively simple and common sense rules. Collectively, the signals systems can be a large fraction of the electrical budget, especially on larger facilities with a high rate of public access and use. Simple and straightforward though they may be, the signals systems are extremely important for the safety and security of the building occupants. This is evidenced by the fact that, where emergency power is available, they are typically so supplied. Otherwise, some other systems such as lighting must be on a backup power source anyway.

Given their importance to the safety of occupants, and their relatively low cost per unit cost, it is advisable to err on the cautious side. If in doubt, provide the extra security on the design documents. Keep in mind that most buildings have a relatively dynamic, changing floor plan and that remodels do not always upgrade the signals systems or even have the requisite funds in the budget. Consequently, if there are extra devices, convenient junction boxes, and nearby home runs, it is more likely the signals will be kept up to code standards in the remodel. A little extra money to build flexibility into the system when a building is built enhances the safety of the facility throughout its useful occupancy.

Exit Signs

The location of exit signs in smaller commercial projects is usually self-evident. The exit corridors and doorways are easily identified as the preferred egress route. Every occupant who enters the passageway must, upon looking right and left, be convinced where the closest exit is and to proceed there with dispatch. Upon arriving at the sign itself, if the location is not itself a door to outside, a large arrow on the sign must indicate where the exit route continues.

The idea of exit signs may seem superfluous for small buildings, especially if the occupants would normally be expected to know their way out without any help at all. However, the signs are for use during fires, when people become disoriented, panic, and sometimes stampede. The building will be full of smoke, it will be dark because the lights are out, and if a fire alarm is blaring, it will be quite noisy. This combination of circumstances can rattle even the most level-headed individual.

The exit signs are designed by strict standards to provide sufficient light through smoke to be readily identified. Only UL listed devices should be used and never devices that do not have an integral light. The codes do allow illumination from a nearby ceiling fixture with battery backup, but why take the chance? The lamps could be burned out, the battery exhausted, or the fixture relocated altogether by a remodel project. Enough things can go wrong with the exit sign already. Depending on an external light source only compounds the possibilities—and the potential liability.

Many spaces other than egress hallways require exit signs, such as large conference rooms or auditoriums. Even large shop or assembly areas need prominent exit signs in view of all occupants. Usually the signs are at ceiling height of 8 to 9 ft and are either ceiling or wall mounted. This same height is appropriate for spaces with a higher ceiling.

For most small commercial projects, the exit signs are provided with an integral battery backup, similar to that used for emergency lights. Where an emergency power source is available, the exit signs should be circuited to this source if possible. This circuit should be clearly identified on the building plans, just as the fixture schedule should require battery backup where required. Some facilities will have both types of exit signs if it is more economical to supply the emergency power to only a limited percentage of the building.

Telephone and Cable

The location of telephone and computer networking outlets is probably the most important information that the electrical designer must solicit from the owner. (The location of duplex outlets is important too, but most designers can locate them at appropriate locations if the space use is known.) As soon as a floor plan is approved by the owner, tentative locations of telephone and computer outlets should either be solicited from the owner or submitted by the designer for review. See Fig. 5.1 for some typical symbols to be used.

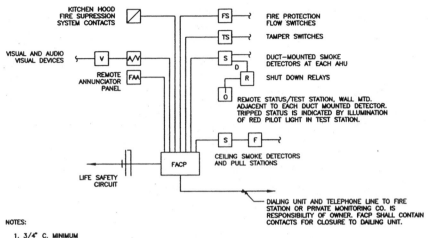

Figure 5.1 Telephone/data symbols.

Some clients may need a little coaching in the use of telephone or computer outlets. Although they may not be needed for current operations, the designer should patiently explain the many potential uses for communication devices. Some uses for telephone lines are for fax machines, computers linked to a local area network or to the Internet, and building controls rooms. Computer outlets may also be needed at individual workstations, dispatch offices, administrative spaces, conference rooms, copy machines, and even fax machines.

The need to identify any current as well as future telephone or computer jacks is important because they are quite easy to install when the building is constructed. The cost does not have to be excessive either. All that is needed is a single junction box with ½- or ¾-in conduit leading to above the ceiling or to the nearest accessible location if the room has a hard gypboard ceiling. With a blank cover on the junction box and a pull string in the conduit, the installation is ready to be upgraded to a communications outlet. It is either this nice custom installation or bare cable hanging down from the ceiling. The latter installation may actually end up costing more anyway. The labor rates (and the time required per installation) are quite a lot less for an electrician than a network certified installer. So the convenience of built-in outlets is well worth the trouble because they reduce the time the network technician needs to install a new system.

Another popular option is to provide device plates with several types of jacks, for example, one telephone and one computer jack. Even if only one is used at the time of installation, the other is available for immediate use once the cable is pulled. The only proviso is that power circuits and communications wiring cannot be in the same electrical feeders. Although it is a very good practice to provide a duplex outlet at each communications outlet, they must be in separate boxes and home runs.

Most projects are designed so that the electrical subcontractor installs the junction boxes, conduit to above the nearest accessible ceiling, nylon or wire pull cord (with plastic grommets at the end of the conduit to protect wire as it is pulled), and cover plates. A separate contractor, or the telephone or network installer, installs the actual wiring. Communications systems have unique demands and needs, so it is best to let this separate contractor install all elements of the system to be sure everything works together as expected.

The only additional item that the electrical designer may need to provide is a cable tray system above the ceiling. This greatly simplifies the organization of the communication wiring in large buildings that have many telephone and computer outlets. Usually a single run or a loop is installed so that the distance to any given outlet is no more that 10 to 25 ft. Hospitals sometimes require the communication wire to be run in conduit, but most other facilities allow plenum-rated wire and cable to run freely in the ceiling space. A cable tray not only helps to organize the wires, but it also helps prevent kinks and sharp turns and protects the wire from damage by the other trades working in the ceiling space.

Communication Rooms

Generally speaking, in office environments, there should be 2 to 3 times as many telephone rooms as electric closets. The reason for this is that up to eight duplex outlets can be on a single home run, whereas every single phone on a computer/modem outlet requires a dedicated home run to the communications panel. Assuming one phone outlet per 2 or 3 duplex outlets, there will be 2 to 3 times as many home runs in a communications system. So, it is important to have more communications rooms so that the home run distances are of a manageable length.

The standard communications setup is changing. Until recently it has been enough to provide a 4- by 8-ft plywood board (painted with two coats of a nonconductive paint) on one wall of the room, with a convenient ground rod, a duplex outlet nearby, and illumination on emergency power. Increasingly this is inadequate (see Fig. 5.2), especially to install the more advanced communications apparatus capable of handling a multiple phone system common to offices, with paging, voice mail, and other automated functions. The size of the room

Signals Circuiting   69

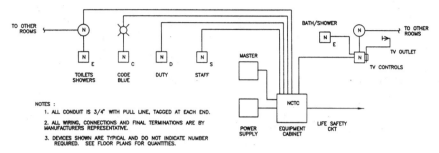

Figure 5.2 Typical telecommunications closet.

doubles, from a 4-ft-wide closet to a room that is at least 15 ft^2. Furthermore, the space requires a dedicated air conditioning diffuser to keep the room heat load within the operating range of the equipment. At a bare minimum, the room needs a thermostat-controlled exhaust fan and a louvered door that can ventilate the space with second-hand cool air from adjacent spaces.

Increasingly, communications is more than telephone and modem connections; it is computer networking as well. Most networks are added to a set of plans almost as an afterthought, and the owner is left to find a suitable location in the office spaces for the file server. A much better solution is to design the network system as a separate communications system, with either dedicated outlets or dual outlets with the telephone connections. A centrally located room should be set aside for the file server, with space for a small workstation to hold the computer, monitor, and an uninterruptible power supply and with shelf space for the software manuals, spare parts, and other related accoutrements. Again, the room should be adequately conditioned so the temperature does not exceed the system operational limits. A single network room per floor or department is adequate because the networking of computers is done by a single loop by which all of the stations are daisy chained to the main server.

Fire Alarm Devices

Fire codes vary with the enforcing authority, but there is a certain minimum of protection that should be provided at any commercial establishment for the safety of the occupants and the property within. (See Fig. 5.3 for a fire alarm riser diagram.) Doing so will often have a short payback, because of lower insurance costs. Even if there are no explicit requirements for some of the following items, it is always better to err on the safe side in issues involving safety:

- A pull station at every exit

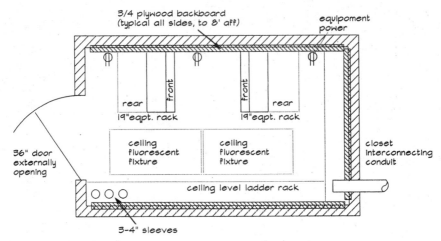

Figure 5.3 Fire alarm riser.

- Fire extinguishers in recessed cabinets at 100-ft intervals in schools, and at appropriate locations otherwise, although always convenient in kitchens and other places with a high fire hazard potential
- Smoke detectors in corridors, restrooms, large common areas, any space with a cooking appliance (e.g., stove or microwave), restrooms, and any space with equipment operating when no one is in attendance (i.e., a network server or mainframe computer that is on 24 h per day)
- Smoke detectors in the supply and return plenums of all air handlers with an automatic fan shutoff interlock and a remote LED test and reset light
- Heat detectors in the mechanical and elevator rooms and any other space containing major electrical equipment
- A main or remote control panel at the main entrance to the building that is easily accessible by fire fighters answering an alarm

In addition to these safety and monitoring devices, it is always good to practice good housekeeping to minimize the potential for fire in the first place. This includes proper storage and adequate ventilation of cleaning materials, chemicals, oily rags, and all other flammable materials. Another common fire hazard is overloading electrical circuits with too many devices. Over time, the high load can cause the insulation around the conductors to become brittle. This can lead to shorts and fires.

ADA Devices

The devices required for compliance with the Americans with Disabilities Act (ADA) are primarily to help assist individuals with auditory impairments to escape a smoke-filled space. This is accomplished by visual strobes that are placed in several specific areas:

- Hallways and lobbies
- Public restrooms
- Large conference or assembly areas of general or common use
- Sleeping accommodations

These devices should be on emergency power, if such is available at the facility. The usual symbol for these strobes is a "V" inside a square box. They must be mounted 80 in above the highest floor level in the space or 6 in below the ceiling, whichever is lower. They must be spaced so that no place in a corridor or other space required to have a visual signal is more than 50 ft from a device, in the horizontal plane. In large rooms more than 100 ft across, the devices can be placed around the perimeter at a maximum 100-ft spacing.

Some requirements for the visual alarms are as follows:

- The system should be integrated into the building or facility alarm system, unless a single-station visual alarm is installed.
- The lamp must be a xenon strobe, or the equivalent.
- The lamp color must be clear or nominal white.
- The pulse duration (i.e., the time between the initial and final points of 10 percent maximum signal intensity) must be 0.2 s, with a maximum duty cycle of 40 percent.
- The lamp intensity must be a minimum of 75 cd.
- The flash rate must be from 1 to 3 Hz.

ADA also recommends audible devices. These should be located in strategic locations to provide additional guidance in finding the egress route and in ensuring the hard-of-hearing get the message. The "A/V" devices, as they are marked on the plans, are situated in egress corridors and in some very large assembly areas such as auditoriums or cafeterias. The audible emergency alarm should be at least 15 dbA above the prevailing sound levels or exceed any maximum sound level by 5 dbA with a duration of 60 s, whichever is louder. The sound level shall in no event exceed 120 dbA.

Other provisions of ADA apply to electrical device locations. Duplex outlets must be mounted 18 in above the finished floor (to the device centerline) for ease of access from a wheelchair. Light and other switches (such as thermostats, ceiling fan controls, lighting control panels, and so forth) must be located 42 in above the finished floor to the centerline, also for ease of access. Similar allowances apply to outlets and switches above counter tops and other less accessible locations.

Otherwise, all operable devices must be within reach of handicapped individuals in a wheelchair. For a space that allows only forward approach, the maximum height is 48 in and the minimum is 15 in. For a clear floor space that allows only parallel approach, the maximum height is 54 in and the minimum is 9 in. These dimensions apply to adults. Heights are less for schools and other facilities primarily used by children. In any event, the electrical and communication system outlets cannot be mounted lower than 15 in above the finished floor.

The ADA has affected other fire and safety devices as well. Exit signs must not only have a specific illumination level but must now have clearly recognizable arrows pointing to the specific exit route.

These are all fairly reasonable requirements that are met with a minimum of extra cost to the owner. Even in areas that do not have to comply with the ADA, they are prudent measures to ensure the safety of all the building occupants. In addition, they provide an extra measure of safety for those who are not disabled, but who might easily panic or become disoriented and would benefit from the additional security.

Computer Networking

Many office layouts are designed and built with no provision for the networking of local computers. There may be one or two staff who use computers but only stand-alone workstations. Even if no one uses a computer in the whole office, it is likely that will change at some time during the life of the building—sooner, rather than later. So the electrical designer does the owner a favor by installing outlets suitable for computer networking. A thrifty and convenient solution is to install cover plates in the wall boxes with two jacks, one for the telephone and one for coaxial cable.

Fiber Optic Cabling

The routing and installation of fiber optic cabling is not often the responsibility of the electrical designer. This is a specialty that is performed as a performance contract by service firms specializing in the area. However, there are several things that the electrical designer can do to facilitate the installation of the fiber optic system.

When a large, comprehensive fiber optic system is installed in a building, the fibers are usually routed in cable trays above the accessible ceiling. The fibers are such that they are quite limited in several ways:

- They are sensitive to electromagnetic fields from motors, air handlers, and other large electrical appliances. The cable trays should come no closer than 3 ft to such equipment. Similar restrictions apply to routing the cable trays near electrical conduits, wiring, lights, and junction boxes.
- The optical fibers have strict limitations on allowable turn radius, which restricts the changes in elevations and direction in all instances.
- Fiber optic cable is extremely difficult (i.e., costly) to splice. Consequently it should be protected from any manner of inadvertent damage by the building occupants, maintenance personnel, and individuals repairing nearby air handlers, motors, ductwork, plumbing, and any other building system.

These special circumstances make it mandatory for the cable trays to be designed and installed in coordination with the other trades. This usually means that the electrical designer must design and route the main trunks and branches in the building to adhere with the above rules. This will make it easier for the fiber optic subcontractor to bid the job (thus resulting in a lower cost to the owner), and will also result in a safer and maintenance-free installation. The important thing is to have the cable trays drawn on the plans and the equipment installed with the rest of the electrical system by the electrical contractor; this will minimize any electromagnetic influence by the installation.

Another important service the electrical designer can provide is a telecommunications closet of the proper size and shape and at the optimum location(s). The size should be of the following dimensions:

Area served (ft^2)	Closet size (ft)
10,000	10 × 11
8,000	10 × 9
5,000	10 × 7

This space must be dedicated to telecommunications service and must not even share space with the electrical installations. There must be one telecommunications closet per floor, and when the area exceeds 10,000 ft (or the horizontal distribution station to a work station exceeds 300 ft), two, or more, closets of the appropriate size are required. Multiple closets must be interconnected by at least one conduit. The closet(s) should be centered in the area served.

Other things that must be provided by the electrical designer for the telecommunications closet (refer to Fig. 5.1) are as follows:

- At least two walls should be lined with fixed $\frac{3}{4}$-in void-free plywood, 8-ft high, and able to support attached equipment. It should be painted with two layers of a light colored nonconducting paint, and there should be a grounding rod near the base of each. In addition, a 1.5-in conduit should be provided from the room for the building main grounding electrode.
- The lights should be mounted at least 9 ft above the finished floor and provide at least 50 fc at a work surface 2.5 ft above the floor. The lighting should be on emergency power or be provided with a battery backup.
- There should not be a false ceiling to the room, but the walls should extend to deck to prevent the intrusion of dust from the ceiling space. The floor, walls, and ceiling should be sealed to reduce dust, and the floor should have antistatic properties.
- Duplex outlets should be placed at 6-ft intervals around the room. In addition there should be two specially marked 20-A, 120-V outlets on separate circuits for equipment power.
- The hvac should be provided 24 h a day, 365 days a year and be of sufficient capacity to handle the entire room load in all seasons. It should be so configured to keep the room at a positive pressure with respect to the rest of the building.

Security Devices

The installation of a basic alarm system is becoming increasingly affordable (especially when reduced insurance premiums are factored into the economic analysis) as the devices become cheaper and the control panel more easily configured. Some of the more common devices are

- Motion sensors
- Door and window contacts
- Magnetic door holds that release a door to close
- Glass break detectors
- Magnetic door locks for emergency exits

These are all inexpensive mechanisms that can be easily installed and coordinated via a small electronic control panel. The panel itself can have its own uninterruptible power supply, and the remote devices are wired via low-voltage wiring. Often the panel has a modem that can signal the security offices for a silent alarm, or an audible alarm can be triggered for local effect.

Many installations combine the security, the fire alarm, and sometimes even the lighting systems controls. Turning all the lights on in a building, or parking lot, in response to a motion sensor or glass break sensor can be a real deterrent to potential interlopers. Likewise, the ability to have some control over lighting can also be a benefit to fire fighters as they evacuate personnel and direct their own people safely to trouble spots on the premises. Combining all the systems in one package also reduces the cost of each system, so that collectively the central automation is more affordable.

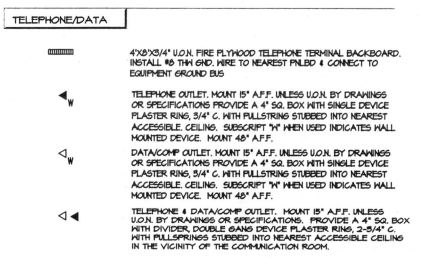

Figure 5.4 Telephone/data Symbols

Chapter 6

Lighting and Circuiting

Light selection and circuiting is one of the few areas in which the electrical designer can affect the visual imprint of a facility. There is a little bit of an art to it, although not requiring much imagination and creativity, if only a few basic principles are followed. Most designers devote minimal attention to the selection and location of fixtures, spending the balance of their design time on calculations and circuiting. They end up with a solid system that works well but is somewhat generic in aspect.

The key to establishing a reputation in lighting design is to be able to spend more time on fixture selection and lighting layout. Finding a good zonal cavity calculations program is the first time-saving step. Most of the programs on the market are so complex and time consuming that designers end up doing all the calculations by hand—or not at all. A quality program reduces the process down to the basics. The program needs only a fundamental set of fixtures and lamps. One that can do batch calculations is a bonus so that when room functions or dimensions change during the design phase, the database can be easily modified and updated. A good program not only saves time but is also something the designer will gladly use. A designer confident in the lighting levels will then design the actual layout with the confidence to be original where it counts.

Another key to introducing some creativity into the lighting design process is to compartmentalize the process. Instead of having total freedom of selection and layout through the whole design process, small tasks are done one at a time, each with only a few variables. If each step is done with care, making simple design decisions at each stage, the final product is not only consistent but solid where it needs to be simple, and creative where it can be.

A solid software program that has all of these attributes, plus an energy analysis routine, is available free from the author. Program features and ordering instructions are at the end of Chap. 13. The following design sequence is molded on this software, although it is generic enough to be applied to any circumstance.

Luminaire and Lamp Selections

Once all the room information has been taken from the plans and input to the calculations program, it is time to think about light selection. Most commercial designs have only a few fixture types. They always have 2 × 4 lay-in fluorescent fixtures, and that is used as the standard. Any other fixture types are considered specialty items and are individually designed. Thus, the initial plan is for 2 × 4s everywhere, usually lensed unless the owner confirms a budget for parabolic fixtures (which are substantially more costly).

The next step is to complete the calculations to determine the number of fixtures required in each room. The space function is known, so the footcandle level can be determined from standard tables. Some representative values are given in Table 6.1. This is the latest standard, which has modifiers for age, contrast required on the work surface, and importance of the task. A general description of these criteria is provided in Tables 6.2 and 6.3. In most buildings only very general assumptions can be made about the age of occupants and type of work to be done. The standard can be used for specialty areas such as classrooms for a specific age group, detailed drafting spaces or print shops, or in areas with a lot of work done on the computer.

With a limited budget, it is good policy to use plain fixtures where possible, so more costly ones can be used in the more visible, public areas. At this stage of the design process, it is okay to use plain lensed 2 × 4 fixtures everywhere. There are two important variables yet to be considered: lamp type and number of lamps per fixture.

Selection of lamp type is usually not a problem. High-efficiency T8 lamps are becoming the standard for most applications. This type of lamp gives a good broad range of light, has a small lumen depreciation factor, and is now manufactured in sufficient quantities to make them cost competitive.

The choice of two-, three-, or four-lamp fixtures is a little more complex. Generally, any area with a footcandle rating of 30 and below should be lit with two-lamp fixtures, up to 70 fc with three-lamp fixtures, and four-lamp fixtures for spaces with higher light levels. These assume a 9-ft ceiling. With a 10-ft or higher ceiling the higher lamp fixtures are needed.

It is good policy, when in doubt, to select a fixture that provides more light than the design target, rather than less. This is especially

TABLE 6.1 Commercial, Institutional, Residential, and Public Assembly Interiors

Area/activity	Category	Area/activity	Category
Auditoriums		Reading	
Assembly	C	Xerograph	F
Social activity	B	Xerography, 3d generation	D
Banks		CRT screens	B
Lobby (general)	C	Ink jet printer	D
Writing area	D	Keyboard reading	D
Teller's stations	E	Machine area	C
Barber shops & beauty parlors	E	Handwritten tasks	
Club and lodge rooms	D	#2 pencil & softer	D
Conference rooms	D	#3 pencil	E
Dance halls & discotheques	B	#4 pencil & harder	F
Drafting		Ball point & felt tip pen	D
High-contrast work	E	Chalkboards	E
Low-contrast work	F	Printed tasks	
Blue line & blueprints	E	6-point type	E
Sepia prints	F	8- & 10-point type	D
Food service facilities		Glossy magazines	D
Cashier	D	News	D
Cleaning	C	Typed originals	D
Dining	B	Residences	
Kitchen	E	General lighting	B
Hotels		Dining	C
Bathrooms	D	Grooming at mirror	D
Bedrooms	D	Ordinary workbench hobby	D
Corridors, elevators, stairs	C	Difficult workbench tasks	E
Front desk	E	Critical workbench tasks	F
Lobby (general)	C	Easel hobbies	E
Lobby (reading)	D	Ironing	D
Libraries		Kitchen counter	E
Active book stacks	D	Kitchen range	E
Inactive book stacks	B	Kitchen sink	E
Cataloging	D	Laundry room	D
Card files	E	Music scores	
Circulation desk	D	Simple	D
Merchandising spaces		Advanced	E
Alteration room	F	Substandard-size scores	F
Fitting room	F	Reading	D
Locker rooms	C	Prolonged, serious	E
Stock rooms, packaging	D	Sewing (hand or machine)	
Offices		Dark fabrics	F
Audio-visual areas	D	Light to medium fabrics	E
Lobbies, lounges	C	Occasional high contrast	D
Mail sorting	E	Table games	D

TABLE 6.2 Illuminance Values for Generic Types of Activities in Interiors

Type of activity	Category	Footcandle range
Public spaces with dark surroundings	A	2-3-5
Simple orientation for short temporary visits	B	5-7.5-10
Working spaces where visual tasks are only occasionally performed	C	10-15-20
Performance of visual tasks of high contrast or large size	D	20-30-50
Performance of visual tasks of medium contrast or small size	E	50-75-100
Performance of visual tasks of low contrast or very small size	F	100-150-200
Performance of visual tasks of low contrast and very small size over a prolonged period	G	200-300-500
Performance of very prolonged and exacting visual tasks	H	500-7650-1000
Performance of very special tasks of extremely low contrast and small size	I	1000-1500-2000

true if some sort of control is to be provided, such as dimming or multiple-level switching. Otherwise, it is a judgment call. A good rule of thumb is that if the lower lamp selection provides light (with the lumen and maintenance depreciation factors included) within 10 percent of the target value, that selection is adequate. Usually all the lamps will not be the same age, so their combined depreciation value is on the order of 60 percent—which will be only a 6 percent variance from the target value.

The Reflected Ceiling Plan

With the computer printout in hand, it is time to evaluate the selections with regard to the actual room layouts. Generally speaking, square-shaped rooms have either one fixture in the middle or two parallel fixtures. Long rooms have the fixtures running lengthwise. In either case, it is always important to remember that more light exits the fixture along its long axis than along its short dimension. That's why it is best to orient 2 × 4 fixtures lengthwise in corridors.

The ideal way to design the lighting is for the electrical designer to locate the lights and to draw the Reflected Ceiling Plan (RCP) when a lay-in ceiling is in use. This gives the engineer the opportunity to orient the lights properly and to space them so as to provide a uniform level of lighting in a space. If the architect does the RCP, the options are limited and unnecessary coordination time is wasted.

TABLE 6.3 Determining Illuminance Values within Categories

General lighting throughout room

Average age of occupants	Room surface reflectance (%)	Illuminance categories		
		A	B	C
Under 40	Over 70	20	50	100
	30–70	20	50	100
	Under 30	20	50	100
40–55	Over 70	20	50	100
	30–70	30	75	150
	Under 30	50	100	200
Over 55	Over 70	30	75	150
	30–70	50	100	200
	Under 30	50	100	200

Illuminance on task

Average age of workers	Speed or accuracy*	Reflectance	D	E	F
	NI	Over 70	200	500	1000
Under 40		30–70	200	500	1000
		Under 30	300	750	4500
	I	Over 70	200	500	1000
		30–70	300	750	1500
		Under 30	300	750	1500
	C	Over 70	300	750	1500
		30–70	300	750	1500
		Under 30	300	750	1500
	NI	Over 70	200	500	1000
40–55		30–70	300	750	1500
		Under 30	300	750	1500
	I	Over 70	300	750	1500
		30–70	300	750	1500
		Under 30	300	750	1500
	C	Over 70	300	750	1500
		30–70	300	750	1500
		Under 30	500	1000	2000
	NI	Over 70	300	750	1500
Over 55		30–70	300	750	1500
		Under 30	300	750	1500
	I	Over 70	300	750	1500
		30–70	300	750	1500
		Under 30	500	1000	2000
	C	Over 70	300	750	1500
		30–70	500	1000	2000
		Under 30	500	1000	2000

*NI = not important; I = important; C = critical.

Some other helpful hints in lighting layout are as follows:

- Do not locate a fixture in the radius of the door swing.
- In commercial office buildings, try to use the same fixture (usually a three-lamp fixture) everywhere so minor office remodels can be done without purchasing new fixtures for the job.
- If significant remodels are expected, specify modular fixtures with a long pigtail, allowing lights to be moved between spots in the ceiling grid without any major rewiring required.
- For major lighting retrofits it is better to replace the entire fixture with a prelamped packaged fixture than to replace the lamps, ballasts, and lens. The cost of the latter may be a little less, but the photometrics and the useful life expectancy of the former are far superior.

Some enhancements can be done in public areas to improve the lighting layout for the public visibility. Some economical ideas are

- A 2×4 fixture in a narrow hallway is awkward. Using narrower 1×4 luminaries adds some visual appeal, for about the same cost.
- Parabolic fixtures in public areas are always nice, budget permitting.
- Downlights should not be used for general area lighting except in specialty areas. However, they can be quite effective when used selectively, such as in an entrance cove of a large office, at a hallway corner, in an elevator or public lobby, above a water fountain, or as wall washes for wall hangings. It is best to use compact fluorescent fixtures or hid lamps instead of incandescent lamps in such circumstances.
- Cove lighting is not expensive, but the architectural enhancements needed for their installation in the ceiling are. So they are not economical for general area lighting but add a very nice dimension in lobbies, public areas, or conference rooms that will benefit from a little more elaborate lighting scheme.

Another dimension of the RCP is selective use of nonstandard grids or gyp board ceilings. All of the design comments so far pertain to 2×4 lay-in ceilings. Two-by-two grids are economical and add a touch of class, especially to small, intimate spaces. Hard ceilings also have their place, such as in a hallway corner, where fitting grid elements together from the two directions might be awkward, or as the border of a conference or meeting room.

Special Areas

Restrooms are a challenge to lighting designers. The mirrors at the sink counter need good lighting, for one thing. An egg-crate grill with fluorescent strip-back lighting provides good illumination. Sometimes this pattern can be extended for the length of the space, for indirect lighting throughout. A few down lights in the rest of the space round out the lighting very effectively.

Another popular option is to have a 1×4 recessed, lensed fixtures at the mirror counter and one or two 2×4 fixtures (depending on the size of the space) for area lighting. This is effective for shower rooms that have a hard ceiling and that need damp-location-rated fixtures.

Conference rooms are another sterling opportunity for the lighting designer. They usually have a higher lighting budget and require several lighting levels. A good policy is to provide both area and spot lighting using parabolic fixtures and fluorescent down lights, respectively. A combination of switching and dimming can provide many lighting combinations. If items will be displayed on the walls, some recessed wall wash fixtures will be nice, or if it is a large room, track lights are an attractive alternative.

Libraries are another creative lighting opportunity. Direct lighting from a higher ceiling at the stacks is best done by a single row of 1×4 parabolic fluorescent lights. Another more costly option is pendant-hung fixtures with some indirect and some direct light. These create a relaxing atmosphere through diffuse area illumination and targeted direct illumination. They are quite effective in libraries that have high ceilings. Bringing the fixture down to 9 or 10 ft above the floor reduces power consumption substantially. The lights can either be run directly above the aisles or at a 45° angle to the stacks. The latter is not quite as efficient, but it is a more visually pleasing organization. In any case, point-to-point calculations should be done to ensure at least 10 fc at the lowest row of books on the rack.

Incandescent lamps do have their place. Their warm, comfortable ambiance is well worth the extra cost in electricity consumption in some situations. Incandescent spots for wall-hung artwork or store displays are more inviting than the harsh illuminance of hid lights. Subtle downlights in restaurants or clubs are a relaxing relief from the impersonal light of fluorescent fixtures.

In any of these situations, the designer should be cognizant of the daylighting available to a space. If the space is in use only during the day, the effect of any special lighting will be greatly diminished. For example, the careful design of spot lights or cove lighting in an exterior hallway will only be effective an hour or two each day. On the other hand, if the hallway is for a theater or concert hall, such lighting can be quite appealing.

Last of all, there are large open storage or public assembly areas such as gymnasiums. The most efficient light sources for such installations are low- or medium-bay lights. Fluorescent lights in a 20- or 30-ft ceiling are very inefficient and also hard to maintain and access.

Circuiting and Switching

Now that the light fixtures have been selected and located, it is time to circuit them. This is not always as easy as it seems, especially when special switching is required. A productive way to circuit the lights is to first draw the circuiting for each room and then connect the blocks of fixtures together to build the complete circuits. There are several reasons for this. First, if the floor plan or ceiling grid is changed in individual spaces, the entire circuit scheme is not affected. Second, going by the circuiting principles in Chap. 7, it is easier to create an efficient, effective design if the fixtures in each room are designed beforehand. Finally, it isolates separate design tasks into discrete, simple objectives that can be done more expertly if the overall building scheme is temporarily ignored.

The design method of choice is to draw junction boxes (with a maximum of four or five connections each), with solid power wiring and dashed switch legs shown between the boxes (Fig. 6.1.) This is a thorough, easily understood and modified system. It is also easy to do with CAD drafting, about as simple to do as showing tick marks for power, neutral, and switch legs as some designers do. Drawing the entire switch leg with dashed circuit lines leaves no room for doubt or interpretation when the system is installed. It also makes the lighting scheme more comprehensible to the owner, the project architect, and bidders.

The general rule for switching is to provide one switch at every space entrance, within reason. For example, if there is a short 10-ft section of hallway with a connecting door on either end, a single switch that can be reached from both entrances will suffice. Generally, all major entrances or exits should be covered so that it is convenient as a person leaves the space from any exit to extinguish the lights (or turn them on easily as the space is entered).

If there are two switches for a single space, three-way switches are necessary, as illustrated in Fig. 6.2a. If three switches are used, the first two are wired as three-way switches, and the third is a four-way switch. This scheme is shown in Fig. 6.2b. If more than three switches are needed for a space, it is best to split up the lights on each circuit. For example, in a very large space half the lights can be on one circuit and half on a second one. The lights in each section should be switched

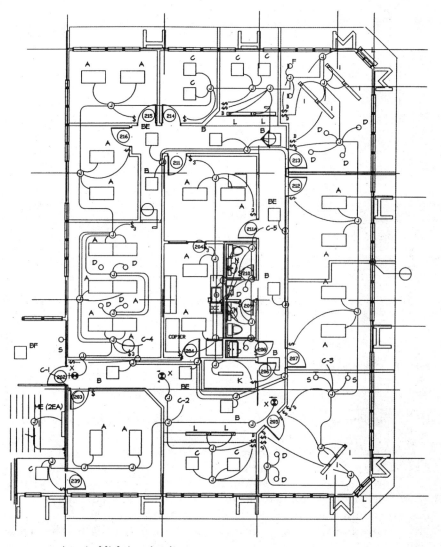

Figure 6.1 A typical lighting circuit.

at appropriate points that are coincident with the area illuminated, with switches for each space at a convenient spot where they intersect.

A little more elaborate scheme is popular with three-lamp fluorescent fixtures. The center lamps are on one circuit, and the outboard lamps are on a second one. This allows three discrete light levels. There is some incremental cost involved in the wiring and an extra door switch, but the energy savings and occupant comfort can more

86   Chapter Six

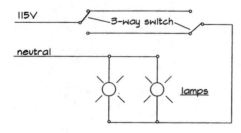

(a) two 3-way switches

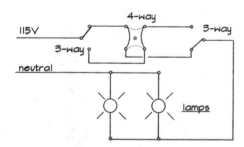

(a) two 3-way switches & one 4-way switch

Figure 6.2 Switching wiring diagrams.

than compensate for this. The application is ideal for multiple-use rooms and for any space with daylighting.

Dimming controls are a little harder to justify but are getting more economical all the time. One possible alternative applies to a row of offices on one exposure of a building, with the same window area and space use. A single dimmer on one of the lighting circuits in each office (assuming they are all wired as detailed in the previous paragraph) can dim the lamps according to the ambient light. The other switch leg is still available for occupant control.

Motion sensors have ideal applications too. Large spaces that are used infrequently are the best candidates. Auditoriums, mechanical rooms, cafeterias, and conference rooms should be considered for motion sensors. Other spaces may be obvious applications too, once the space is in normal use. The door switch can be changed out to a combination motion sensor-switch for effective control in such a circumstance.

Photocells should be a standard for all exterior lighting, in combination with either a regular switch or a time clock. A useful indoor application is for large spaces with a wide expanse of glass that intro-

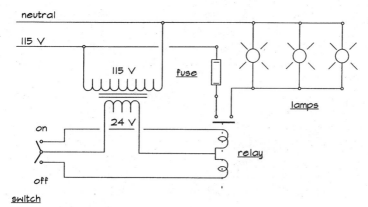

Figure 6.3 Remote control circuit.

duces daylighting much of the day. If the exterior 10 ft of area lights are on a separate switch leg, these lights can be controlled by a photocell or even a dimmer for substantial energy savings and better occupant comfort.

A useful feature for some applications is to have a pilot light in series with the switch. This is helpful for spaces that are rarely used, such as mechanical rooms, janitors closets, attics, or pipe basements. If the occupant leaves the room and forgets to turn out the lights, the pilot will be a reminder to turn them off.

Finally, large spaces with many different lighting zones such as a library, assembly hall, or auditorium make special demands on the switching. Often a lighting control panel is the most economical solution because the switch leg for each circuit does not have to be brought all the way back to the switching point. Each circuit is controlled by a low-voltage loop with a relay at the lights controlled, similar to the diagram in Fig. 6.3. The control panel has many programmable features, including dimming, time-clock functions, and interfacing with a local or remote photocell or motion sensor. Another application for remote control circuits is site lighting, where the switch legs can be quite long.

Conserving Energy

Several energy-saving practices have already been elucidated as an important part of the design process. The use of high-efficiency fluorescent lamps and electronic ballasts should almost be a standard on any design job. The savings so outweigh the small additional cost that this equipment should be used unless there is a compelling reason not to use it.

One innovation to the standardization of an entire facility on the T8 lamps is to have three standard lamps. The entire design process is done as described already, using the most common lamp, usually a 32-W T8 lamp. Once all the calculations are done, two more high-efficiency T8 lamps of lower wattage and lumens rating are introduced to the selection process. A computer program can be used to automatically review each room and select the lamp that most closely approximates the target footcandle level. Because up to 50 percent of the rooms will have been overdesigned, using one of the other lamps will reduce the energy consumption. This procedure in a large office complex can reduce lighting costs by up to 15 percent, with a minimal extra hassle for the maintenance personnel.

Electromagnetic Theory

There are several interesting aspects of the physics of electromagnetic charges that transfer to lighting design in a very practical way. An awareness of these general principals can be helpful to lighting designers in their understanding of the phenomena and in the application of the latest technology.

The most fundamental concept of lighting is that the intensity of light diminishes as a function of the square of the distance from the source. This is easy to remember if you think of the source as a point source, or small intense sphere of light. If you move away from this source, the area of the sphere increases as the square of the radius of the sphere. This is true even if the light is focused or otherwise concentrated at the source—you usually cannot escape the relationship.

A significant change to this effect occurs when investigating the illumination from an infinitely long line of lights (or, at least, an approximation thereof). In this instance the light level diminishes as a direct function of the distance. If you are 10 ft from the line source, the illumination drops only to 10 percent of the source, not 1 percent of the source intensity as in the case of point sources. This effect can be put to use in long hallways with cove lighting by a single row of fluorescent lamps to produce a very uniform distribution of light in the hallway.

Another interesting electromagnetic relationship applies to the light emanating from an infinite flat surface or reasonable proximity thereto. Here the intensity of light is the same at all distances from the source. A luminous ceiling, for example, will provide a consistent light level no matter the distance from the ceiling if the room is sufficiently large and the distance from the ceiling is sufficiently short by comparison. This effect can be used in spaces such as laboratories, operating theatres, or computer chip assembly areas, where the quality of lighting is critical to the task or activity.

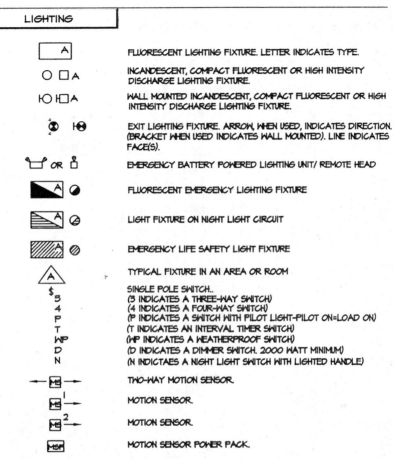

Figure 6.4 Lighting symbols.

LIGHTING FIXTURE SCHEDULE

FIXTURE DESIG.	FIXTURE DESCRIPTION	LAMPS NUMBER/TYPE	VOLTAGE	MOUNTING	REMARKS
A	2'X4' PARABOLIC FLUOR. FIXTURE LITHONIA 2PM3-F-B-332-18LD-120-GEB10	3/32V/T8/SPX35	120	RECESSED	
B	2'X2' PARABOLIC FLUOR. FIXTURE LITHONIA 20PM3-F-B-217-9LD-120-GEB10	2/17V/T8/SPX35	120	SUSPENDED	
C	2'X2' PARABOLIC FLUOR. FIXTURE LITHONIA 20PM3-F-B-417-9LD-120-GEB10	4/17V/T8/SPX35	120	RECESSED	
D	HORIZONTAL BURNING COMPACT FLUOR. DOWNLIGHT LITHONIA CF-2-11-2/18DTT-84A-120-GEB10-EL	DTT/26V	120	RECESSED	PROVIDE ELECTRONIC DIMMING BALLAST (where specified on the plans)
E	ARCHITECTURAL BOLLARD LITHONIA KBE8-70S-R5-120-H36-DDB	70V/HPS	120	FLOOR	
F	VERTICAL BURNING COMPACT FLUOR. OPEN WALLWASH LITHONIA AFV-6-32TRT-6AR-LD-120	DTT/26V	120	RECESSED	PROVIDE ELECTRONIC DIMMING BALLAST
G	UP/DOWN GENERAL ILLUMINATION FLUOR. FIXTURE LAM-GCA-2T8-X-X-120-EB	2/32V/T8	120	RECESSED	
H	SITE LIGHTING LITHONIA KL-70S-A-120-BBP-DDB	1/70V/HPS	120	POLE MOUNTED	
I	DIRECT/INDIRECT FLUOR. FIXTURE EXECULITE EXC5I2-B-4/1-T8-120-NS-S18-BV-3' (4")	5/F32/T8	120	RECESSED	PROVIDE ELECTRONIC DIMMING BALLAST
J	EXTERIOR AREA LIGHTING LITHONIA KFL2-150S-70S-120-V3-SF-FV-DDB	1/150W/MH	120	WALL	
K	GENERAL PURPOSE FLUOR. FIXTURE LITHONIA AV-2-48-120-HRUN-VGAFPV	2/32V/T8	120	SURFACE	PROVIDE ELECTRONIC DIMMING BALLAST
L	FLUOR. COVE LIGHT ELLIPTIPAR F-301-A332-S-00-120-000	2/32V/T8	120	SURFACE	LOCATE IN ARCHITECTURAL COVE
M	CORRIDOR WRAPAROUND FLUOR. FIXTURE LITHONIA CB-240-A-120-GEB10	2/32V/T8	120	SURFACE	
N	COMPACT FLUORESCENT STRIP LIGHT LITHONIA S-1-CF40-120-SSB	1/T5/40V/RS	120	CLNG, SURFACE	
S	WALL SCONCE	DTT/26V	120	WALL	
X1, X2	CEILING MOUNTED LED EXIT SIGN 1 OR 2 FACES LITHONIA FRP-V-L2-RC-DA-120-ELN	F8T5/14V	120	CLNT, SURFACE	UNSWITCHED

NOTES: 1. SUFFIX "E" INDICATES A STANDARD NICAD LITHONIA BATTERY PACK IS REQUIRED.
2. SUFFIX "F" INDICATES A ONE HOUR RATED ENCLOSURE SUPPLIED BY MFGR.

Figure 6.5 Lighting fixture schedule.

Chapter

7

Power Distribution System

The design of the overall power distribution system deserves much more attention than it usually receives. Typically the architect assigns to electrical a small space in some remote corner of the building plans. It is rarely large enough for the service entrance and main distribution panel, so a lot of energy must be devoted to lobbying for a larger room. Little energy or enthusiasm remains for an even more important consideration: the location of the room to begin with. If that's not enough, rare is the architect who allocates space for a dedicated electrical closet anywhere in the plans. The panels, transformers, and other electrical gear usually end up in mechanical rooms, closets, or even in hallways.

It is up to the electrical designer to be assertive in the early, schematic phase of the project. That is the time to negotiate for a large enough room for the main service gear and a few remote rooms for the local panels. The compelling reason to do so is significant: cost. No other engineering discipline uses up as much of the construction budget as electrical, so it must be made clear that this nominal extra consideration is well deserved.

Operations Research

There is a very complex and sophisticated realm of modern statistics and mathematics called *operations research*. It is a planned, logical approach to what might seem to be a very simple problem: given a set of nodes and specific allowable lines between them, what is the shortest distance from one point to another. One practical example of this is the routing of interstate commerce across the country.

Imagine a map of the United States, and each major city is a node and the roads between them are the allowable paths. Operations research develops a method that defines the optimum path to go from one point to any other destination. It does so by defining a value to each line. At first you would think this value might be just the distance between the nodes on either end. However, in the case of moving goods by roadway, the distance is not as important as the time. Or, if you really want to get to what is most important, the cost. This is a combination of time and distance, because distance costs in terms of both time and cost, for fuel, wear on the equipment, increased maintenance cost, and even tolls on some roads.

The same process takes place every summer vacation, with the family stretched out in front of a map of roads that go in the general direction of Disneyland, the mountains, or some other favored destination. There are multiple possible routes, and more often than not the final route is not the shortest distance between home and Disneyland. A detour is made to a larger city where overnight rates are cheaper or to a state park where camping is inexpensive. If a lot of time must be made up between destinations, super highways are taken; otherwise smaller country roads are the choice, to add intrinsic value in the guise of variety and interest to the trip—and pleasant road stops to picnic for lunch or dinner.

Operations research simply quantifies this whole process. The objective in this case is to maximize value, as defined by a combination of cost, time, interest, and efficiency. When the whole process is put into terms of mathematics, it quickly becomes a very complex process indeed once all of the important factors are accounted for. Yet, all of these machinations go on in the mind of the average summer vacation planner, and the solution becomes in most cases quite a good approximation to the mathematical solution.

The important thing to recognize in this whole process is that whenever a choice has to be made, there are not only many possible choices but many ways to quantify them all, too. The family planning a vacation has feelings that quantify the routes, and it is up to the parents to take the decision of least resistance between everyone's conflicting needs and wishes.

The Design Team

There are many such decisions in the design of an electrical system, too. These decisions can be precisely analyzed by the methods of operations research; yet, just like the planning of a vacation trip that is done very capably with virtually no knowledge of the mathematics of the whole analysis, the planning of an electrical system can benefit just as much from some careful planning. The important thing is to be able to quantify the alternatives—in this case the routes between the nodes, or points of use.

This is really not such a hard thing to do. First define what it is that we are quantifying. The routes are the home runs, and a reasonable way to think of the "nodes" is as the average distance of the home run to the panel. Next, consider the "values" that are relative to each of these nodes:

- The cost of the materials: conduit and wiring.
- The cost of the labor: to route the conduit and pull the wire.
- The number of junction boxes required by code or convenience.
- The difficulty of the routing and conflicts with the other disciplines; that is, can a journeyman do the job or is a master required?

The experienced supervisor can probably take a short look at the plans and make all of these important decisions intuitively. She or he knows whom to assign to what task and how long it will take. These field hands, though, are limited by the design itself—the panel has already been located, the nodes defined, and the home runs designated.

It is a mathematical axiom that the fewer the unknowns in an equation, the easier it is to arrive at a solution and the more exact that solution will be. It goes without saying that there are many, many unknowns in this situation, even for a small building. It is also logical to say that had the designer taken the time to more carefully lay out the panel, home runs, and other elements of the power distribution system, there would be fewer unknowns for the contractor to solve in the field. Thus, it is a simpler problem to solve and one that could therefore be done quicker, cheaper, and more accurately.

This is the name of the game, is it not? To be able to do something with less time, materials, and trouble means everything to a contractor. An engineer who develops a set of plans that has these special attributes will have lower bid prices, less trouble in the field, and a happier and more productive contractor. How, then, to create such a work of art?

Decision Making

It is difficult to have any respect for the electrical engineer who, and this is fairly common, lays out the power plan only haphazardly, on the understanding that "the contractor will do it all over again anyway." This is a lazy, irresponsible attitude. The fewer decisions that the contractor must make (and no matter how thorough or accurate the design, a great many choices always remain to be done on site), the better the decisions that will be made. There must be a very clear distinction between the two, and it is the engineer's responsibility to define that line by requiring that all points of use be located exactly where shown and circuited as illustrated. The result will be a better job, less potential for liability, and less trouble in the field.

The contractor is expert at pruning the trees; the engineer must have the whole forest in mind. This implies several things that the engineer must accomplish, first illustrated in the case of designing the low-voltage power system:

- The quantity and location of all duplex outlets should be exact.
- Each home run should point toward the panel location and should start at the circuit element closest to that panel.
- The location and use of all special outlets (i.e., refrigerator, Xerox machine, or GFI restroom outlets) should be identified.
- The composition of the elements on each circuit should be carefully determined.

Once this is done, the center point of each home run can be determined. Marking each spot clearly with a red marker can designate the node of each circuit. Then the panel(s) can be located so as to minimize the distance from each node to the panel.

A less exact version of this process is to situate the panel in the geographic center of the space. This works well for a building of a consistent floor plan, such as all-office spaces. More complex designs done in this way can be a problem. If half an area is made up of only classrooms and the other half contains small offices, the "center of power" is closer to the center of the offices than to the center of the whole space. Locating the panel at the power center would logically minimize the average conduit run to the panel.

It is important to locate all power uses as single nodes because everything contributes to the power center. This includes normal duplex outlets and dedicated, special outlets, too. The effect of these special outlets, or concentrations of power, is to shift the optimum location of the panel off-center, in the direction of the concentration. By the same token, if there is a lower concentration of power use to one side, as might be due to the location of several classrooms there, the logical panel location will move away from the classroom side of the building.

Even if all of the outlets and home runs have not yet been located, a good estimate is possible using a chart of recommended power densities. Although this table is intended as a guide to energy conservation by providing only those outlets that are necessary, and no more, it is also a good guide to the relative count of outlets in a variety of spaces, as might occur in a hospital (see Table 4.5).

The Big Picture for Power

All of this may seem to complicate an otherwise simple process. By quantifying the analysis, however, it can be extrapolated for use in any situation. Once all of the points of use are identified on a plan, several schemes can be tried out. It may be more economical, for example, to provide a small panel in a high-power-density area with a feeder to the main panel. What was once a judgement call that could only be made by the most experienced estimator—thus a very valuable individual in any organization, whose time has many demands—can be made almost as accurately by someone less qualified.

The difference in conduit costs is easy to estimate, once the center of power served by the panel is determined. If several home runs are to be combined into a single conduit, the best place to begin this conduit run is at the center of power of the respective home runs. The length of this larger conduit is the distance from that center to the panel. This can be estimated with confidence, and the reduction in rating required of the wires is determined just as carefully.

The same technique can be used to situate the main distribution panelboard (MDP) with respect to the remote panels. Using a small-scale plan that shows the whole floor, the panels become the nodes and the MDP is situated as nearly equidistant from them all as possible. There will probably be different amp ratings for these panels, and because the feeder to higher-amp panels is more expensive and harder to run, such panels need to be given a greater weight in the analysis. Drawing a circle around the point node can be used to designate a higher-amp panel. The impact on locating the MDP will be such that the center of power is shifted toward the high-amp panel, to shorten the feeder run.

Of course, everything is only hypothetical until the architect actually assigns the space for the equipment. If some means of reasoning can be presented, though, other than a blanket request, even architects will be accommodating. Providing some insight into the impact on cost can make the engineer's arguments even more compelling. The data generated by the basic operations research methods provides both the organization and cost backup and should play a crucial role in gaining the engineer the real estate needed.

Three-Phase Zoning

It is instructive to compare the design process to a similar procedure used by mechanical engineers in the sizing of the hvac equipment. This is a highly developed method that has many of the same driving forces as the electrical distribution system presented. The design method used by the mechanical engineer is

1. Run the building loads to determine the air needed in each space.
2. Segregate specific areas of the building to be supplied by a single control box, such as interior and exterior zones and specialty areas.
3. Arrange the main supply ducts to deliver air to these zones in the most direct and efficient way.
4. Situate the air handler in a central location to minimize the return and supply duct runs.

This method is similar enough to the electrical process that it is not unexpected that often the most expeditious location for electric panels is in the mechanical rooms.

The key to the hvac design is the concept of zoning. Exterior zones have outside walls and a roof and have different heating and cooling needs than interior zones. Consequently, they are isolated on separate systems. Computer simulation runs are used to calculate the air supply to each space in this zone. Electrically, rooms of different occupancy can be thought of as being unique zones in that the power needs and receptacle density varies. For example, an office has different needs than a storeroom or lobby.

The hvac designer isolates all the zones and numbers them according to the data input to the load software. The values assigned by the electrical designer would be the estimated quantity of outlets in each room, and the zones would be areas served by each circuit. The concept of exterior and interior zones has a little different significance and is related to the balancing of the three phases on the panel.

The objective of the electrical design is to match the load on each of the three phases. This minimizes neutral current losses and optimizes the efficiency of the entire power distribution system. There are some simple rules that establish an idealized, if sometimes unrealistic, situation. However, they should be the goal of every conscientious electrical engineer:

- Design the power distribution system three circuits at a time.
- These circuits are connected to the three separate phases at the panel.
- The same occupancy and space use should occur in the area served by each circuit.
- Each circuit should have the same number of outlets.
- If different space types are served, the load should be approximated to be the same; thus, more or fewer outlets should be used in nonstandard zones to match the load on the other phases.

- Allowance should be made for resistance losses in the home runs. Ideally, they would all be the same. If not, the equivalent of one outlet more or less should be added to or subtracted from the circuit to simulate the actual loading condition.

These are some fairly strict requirements that cannot be met unless the load on each and every outlet is known. Such data is rarely known prior to occupancy, but the engineer should make an effort to learn enough about the tenant and space use from the owner to make reasonable judgments.

Rarely is this kind of attention devoted to electrical design. The reason to do so is simple: neutral current. In a simplistic sense, when the three phases are unbalanced, current flows in the neural to balance them, and this current goes to ground and is wasted energy. In the case of the electrical panel the unbalanced current is in the feeder to that panel. So there can be some imbalance between any three circuits connected to the main panel bus. The purpose in the above micro-management of the phases is this: The more closely individual groups of circuits are to having the phases balanced, the more closely the overall balance will be when all of the circuits are considered. Also, the electrician can be instructed in the project documents to balance the phases, which most will at least attempt. Otherwise, the better the phases are balanced on the plan work, the more productive the electrician's efforts on the job.

A little less rigorous way to balance the three phases in the panel is to divide the whole area that will be served from a single panel into three separate zones. For a good estimate of the total power use, multiply the area by the average power density of the space type. Each large zone, comprised of diverse spaces of unique power densities, will have equal power requirements and all circuits will be on a specific phase. This broad criterion guides the overall design process and results in a design fairly close to the ideal, with less trouble. The individual circuits within each zone are designed, if no home runs cross the borders of the respective zones, according to the limiting capacity of the wiring.

This process assumes all spaces are in use during the same hours of the day. Areas that have different occupancy should not be included in the zoning but should be circuited according to the small-space procedures so that when they are in use, the circuits will be relatively evenly balanced.

This is quite similar to the hvac design process. Interior and exterior zones, we have seen, are on separate controls but are served from the same central air handler. Computer rooms or other spaces with unique requirements are on separate air handlers altogether, especially if they require air conditioning 24 h a day. This keeps the central system from operating when the load is very small and the efficiency of the system is at its lowest.

Isolating electrical service to spaces in use outside of the regular operating hours is a good idea, too, especially if a large transformer can be taken off-line. That is, when the power use is nominal, the efficiency of the transformer is low and standby losses are high. Providing a small, dedicated transformer to special spaces that will be fully loaded after hours when the rest of the facility is unoccupied will save a lot of power.

Each of these procedures for reducing neutral current may be bothersome to the design engineer and the electrical contractor. However, their impact on the energy use by the building for its entire lifetime far outweighs any added cost to the design or construction phases of the work. For this reason, if not for sheer professionalism, in honest acknowledgment of the owner's needs, it is important to develop the habit of careful circuiting of all elements of the power distribution system.

Lighting

Many of these same principles apply to the circuiting of lighting. A practical approach is to balance the lighting and not try to incorporate the three-phase analysis with the power circuits. There is a very logical reason for this: Lighting is in use for several hours each day when few or no power circuits are energized. Most lights remain on in a typical office, regardless of occupancy. This may include blocks of an hour or more early in the morning, at lunchtime, late in the workday, and then for the cleaning crews in the evening hours. If the lighting is balanced independently, neutral current will be minimized even during these times.

Other circuiting principles apply, too. For one thing, the home runs need to have the last element in the direction of the panel to minimize the conduit and wiring needed for the run to the panelboard. The lighting can even be zoned according to use patterns so that the load on the panel will be balanced all the time.

The circuits themselves are created by much the same method as prescribed for the duplex outlets and power, with an extra constraint. First, the lights in every room are circuited, using junction boxes and showing the hot and switch legs as solid and dashed lines, respectively. These junction boxes are now the nodes for the system, and they are combined in a very specific way so as to minimize material costs and resistance losses in the home runs. Ideally, each circuit would terminate near to the panel so that the actual home run is very short. Each circuit would fan out from the panel to the most remote areas of the facility.

This is not always possible or practical. Again, it is intended as a goal, not a requirement. The purpose is to get away from the common practice of circuiting small blocks of rooms, each with a long home run to the panel. Often these home runs are combined into a larger conduit, and the amp carrying capacity of the wire must be reduced. Although conduit cost is reduced, wiring sizes are increased for the extent of the whole circuit.

This circumstance can be avoided altogether by simply drawing the circuits so that they all terminate near to the panel. This is not hard to do if the lighting in each room is first circuited with boxes and switch legs. Then room nodes (circuits) can be combined for the shortest possible home runs. Hallway light circuits (which have low amps but cover a long distance) can be used to bring some remote blocks of lights or large specialty areas closer to the panel.

It is not a common design practice for the electrical engineer to show junction boxes. This is usually left up to the electrical contractor. It is recommended here not only to help in the node-circuiting model but also to bring the design a little further along before it reaches the field. There are several things that this accomplishes:

- There is a clear indication, and thus a limit, on the feeders from each junction box. A good maximum is four, and the designer should strive to use all four for every box to make best use of materials and workers.
- The estimators, in using the drawings for their bid prices, will be able to generate their cost faster and with more confidence. This reduces the markup they must put on the price to cover unforeseen contingencies and gets the owner a better price for the work.
- Installation by the contractor will be faster because fewer decisions will have to be made at the job site.
- There will be fewer problems in the construction phase because the decisions that are made will be installation-related decisions over which the contractor has complete control and scope. There is no greater favor the engineer can do for a contractor than to give freedom on the job site.

Demand Factors

Individual circuits are best designed for a maximum 2 percent voltage drop at full load. The fewer appliances that are on at any time, the less current flows and the less power that is lost to voltage drop. This means it is best to design circuits so that several diverse areas are served. Instead of putting all the lights in a storeroom (which are used sporadically and for a short period) on one circuit, divide them among several adjacent areas that are in more constant use. The same principle applies to power circuits. This approach not only reduces resistance losses but duplicates service to important spaces so that when one circuit is lost, another is available.

Demand factors are applied for the total load on a panel. These assume that not all circuits are in use simultaneously, so the panel and feeder ratings do not have to be sized for the full load. These factors are used before calculating the nominal 2 percent voltage drop criteria, so this is a guideline that applies to all circumstances. Another common guideline is to size a feeder for all of the noncontinuous loads plus 125 percent of the continuous load. There are several types of loads: receptacles, motors, and lights.

Receptacle demand factors

For power circuits the first 10 kVA of load is used at 100 percent, and the remainder is used at 50 percent.

Motor demand factors

The general rule for sizing a feeder serving one or more motors is for a capacity at least equal to the total full-load amps of the connected motors, plus 25 percent of the highest-rated motor in the group. In addition, if there are any more items on the circuit, their load must be calculated by their respective criteria and then added to the total. For the motor portion of the load, there is no allowed demand factor.

There are extenuating circumstances to this method, as permitted by the authority having jurisdiction. For example, the motors might operate intermittently or on a duty cycle. Alternatively, they may be configured so as not to operate at the same time (e.g., a backup pump). In these cases, it is reasonable to reduce the size of the main feeder so that the highest expected load is met.

Lighting demand factors

Calculating the lighting load factors is a little more complicated. Table 7.1 lists the general criteria.

Conduit Supports

A reasonable guideline for supporting rigid metal conduit runs is as follows:

Power Distribution System

TABLE 7.1 Lighting Load Feeder Demand Factors

Type of occupancy	Portion of lighting load to which demand factor applies (VA)	Demand factor (%)
Dwelling units	First 300 or less at	100
	From 3001 to 120,000 at	35
	Remainder over 120,000 at	25
Hospitals	First 50,000 or less at	40
	Remainder over 50,000 at	20
Hotels and motels plus apartments with no cooking provisions	First 20,000 or less at	50
	From 20,001 to 100,000 at	40
	Remainder over 100,000 at	30
Warehouses (storage)	First 12,500 or less at	100
	Remainder over 12,500 at	50
All others	Total voltamperes	100

Conduit size (in)	Distance between supports
$1/2 - 3/4$	10 ft
1	12 ft
$1 1/4 - 1 1/2$	14 ft
$2 - 2 1/2$	16 in
3+	20 in

These are maximum support distance spacings for straight runs of conduit. It is best to also provide supports at each elbow, every significant change in elevation, and near permanent equipment connections including junction boxes. Vertical runs should be supported at each floor at a minimum. It is also a good idea to add supports in congested areas, where personnel traffic is expected, and when the conduit runs near vibrating mechanical equipment. Long straight runs in any instance must have flexible supports to allow for expansion and contraction of rigid conduit.

Rigid conduit is the best choice for major feeders, for high-amp or voltage lines, and for any electrical service in areas with the potential for damage. This includes mechanical and electrical rooms and any area where the conduit is exposed or accessible to the public. This provides the maximum protection from damage and for public safety.

The conduit should be run parallel to building lines in a neat manner. It is usually run flush with the structure, leaving room for the other building utilities below. It is best not to run any conduit along a roof because it causes problems when repairing or replacing the roof. It is best to run the service below the roof, up through a curb or pitch pan to a disconnect on a separate unistrut rack, and then via flexible waterproof metal conduit to the unit itself.

The Service Load

A good way to sum up all of the various aspects of the building load is with a standard form, such as that in Table 7.2. This type of form not only ensures that all parts of the load are accounted for but also that the proper factors are used in considering the pertinent demand factors. It is a good idea to include this tabulation on the bid documents because it contains all of the information needed by the permitting authority to properly evaluate the project.

TABLE 7.2 Electrical Load Analysis

NOTES: Building "D" 20,500 ft^2	kW
I. Lighting load	
A. Lighting load ___40.5___ kW × 1.25	50.62
B. _____ kW × 1.25	
II. Receptacle load	
A. 20.5 kW − 10 kW = 10.5 × 50% = 5.25 kW + 10 kW	15.25
B.	
III. Equipment load	
A. AC	87
B. Heat	
C. Water heater(s)	3.80
D. Elevator(s)	
E. Vending machines	2.60
F. Miscellaneous	6.80
G.	
H.	
I.	
IV. Kitchen load	
A. _____ kW × _____ × demand factor (NEC 220-20)	
V. 25% of largest motor	2.20
VI. Total connected load (kW)	169
VII. Total connected load 169 kW/0.9 pf = 186 kVA	
VIII. Future load _____ kVA + total connected load _____ kVA = kVA	
IX Total amp load of 517 A. at 208 V., 3 phase, 4 wire	
REMARKS:	

Chapter

8

Designing the Visual Environment

There is usually not a lot of effort devoted to creating a visually appealing and energy-efficient artificial lighting scheme, especially on smaller projects with a limited budget and design time. Design-build contracts are particularly susceptible to this flaw, but even architects are not known to be overly creative on small jobs. This is unfortunate because there are some very nice things that can be done with lighting at a nominal cost but with a very noticeable aesthetic impact. This chapter discusses a wide range of alternatives.

Maximizing Impact

The first task is to survey the entire lighting layout for the project. Typically the lights can already be located and circuited, using standard 2 × 4 lensed fixtures, industrial turret lights, and other economical if unremarkable luminaires. Having gone through the design and selection process, the lighting designer will be familiar with the uses of each space and the needs of the client. He or she should also have an idea where lights will be least noticed and most utilitarian. Typically low-priority spaces would be storage areas, open office or modular furniture spaces, and hallways that are exclusively for staff use. These spaces can have the most utilitarian, least-expensive fixtures.

Other spaces with a higher public visibility require a higher-quality luminaire. These spaces include main lobbies, public corridors and restrooms, conference rooms, and offices of senior executives. This category might also include some common staff areas such as break rooms and dining halls where a little extra attention will have a strong positive effect upon many people.

The objective is to use available funds in the most effective way possible, even if it means specifying twice as many fixture types. Lower-quality fixtures in some spaces offset by higher-quality ones in others, with little noticeable effect to the occupants of the former, and a marked difference in the latter, is a design well done. This can even be done within the original lighting budget if the designer is willing to devote a little more attention to detail.

Some examples of what these trade-offs might be in a typical office complex are

- Using lensed fixtures in staff offices and parabolic ones in executive offices for a more elegant environment to match the more expensive furnishings
- Installing 1×2 lensed fixtures in staff corridors and 2×2 parabolic fixtures in public corridors to dress them up a little
- Three 4-lamp 2×4 fixtures in a large staff work room and four 3-lamp fixtures (for a more uniform lighting with less contrast) in a large individual office space
- Recessed 2×4 lensed fixtures in a public restroom and surface-mounted wraparound lensed fixtures in a staff restroom (both with hard ceiling)

The idea is that most occupants rarely notice the lights unless they are out or malfunctioning. However, they do notice when special or unusual fixtures are installed in a room. So it is important to identify those spaces where the extra quality will be most needed or appreciated, either by the occupant or important clients or visitors. Finally, if the space will be unkempt no matter what, either due to the activity therein or the work habits of the occupant, there is no need to provide anything elaborate regardless of the circumstances.

This kind of attention to detail is rare in small projects, especially those without an architect involved in the planning. Often it is enough to impress the client if the electrical designer just asks leading questions about the fixture selections without actually having to implement wholesale changes. It gives the sincere impression that the designer is quite concerned about the clients' needs and anxious to create a quality project even if it takes more design time for the same fee.

Regardless of the space priority, the light level should be calculated exactly using the unique space characteristics and the photometric tables of the luminaire. This means more than using the room dimensions and approximate reflectance values. It means using the exact reflectance of the ceiling or acoustic tiles, the floor covering, and the walls. Another important factor is in spaces with many bookshelves, dark furniture, or modular furniture and panels. These spaces will need more lighting than will an open space of the same dimensions. These anomalies do not necessarily have to be accounted for in exact point-to-point calculations, just in a quality zonal cavity computer program that considers all the different factors.

Lamp Selection

Almost as important as the selection of fixtures is the selection of the lamps. The T8 series seems to have superior color rendition and is ideal for classroom applications and many other locations. These lamps are not the only selection for fluorescent lamps, and the client may appreciate the opportunity to choose the optimum lamp. This is particularly important in spaces with critical visual requirements, such as selecting fabric in a retail store, illuminating mirrors in a cosmetic showroom, or lighting layout tables in a magazine advertising department. Such spaces might benefit from cool or warm shades or even lamps that more closely imitate natural daylight. The lamp selection may even take priority over fixture selection.

When such nonstandard lamps are chosen, the lighting calculations must be modified accordingly. Ideally, the exact lamp should be used in the computer analysis program, although the light level can be approximated by proportioning the lumens output of the lamp types. Another consideration is that the photometrics of fixtures change depending on the size of lamp used. A fixture designed for T8 lamps, for example, will have slightly different photometrics than when T12 lamps are installed. Changing lamp sizes may also affect the aesthetics of open fixtures, such as parabolic or other expensive fluorescent fixtures.

Another important rule is to never mix light sources in the same space, unless by careful design. A common mistake is to have several types of exterior light sources: high-pressure sodium for parking lot lights, metal halide for recessed down lights above an entrance walkway, and even incandescent sconces at the entrance itself. Each type of lamp creates a slightly different color of illumination, and it does not look nice when they are mixed together.

Designing the Visual Environment

Most electrical designers scoff at this wonderful opportunity to do a little of the architect's creative work. Most architects do the lighting design themselves, but even they are willing to delegate the task to a competent engineer or designer. There are several reasons why this is a sterling chance that should not be passed up:

- Having the ability to design lighting is a bonus on any resume or for any promotion or job interview.

- A good lighting designer can attract important new clients to a firm because good lighting designers are in high demand and hard to find in most cities.
- Because electrical designers are already familiar with most of the light fixtures on the market, their photometrics, and other characteristics, it is only natural for them to design the lighting as well. Architects have many other things to keep track of during the preparation of construction documents.
- Lighting design is the only aspect of the electrical service that permits the use of creativity and that has a significant impact upon the building environment.

These are some important incentives to develop a repertoire of lighting schemes, to be used selectively, on even the lowest-budget projects.

Identify a Fixture Budget

The key to a fine lighting design on any project is to establish a firm lighting budget. It is important to do this, in writing and with the owner's approval, very early in the design development—or even the schematic—phase. The reason for this is that lighting is always the first target for cost reduction on any project overbudget. By establishing a reasonable budget first, when it comes time to bargain, the lighting will be much farther down the list of cutbacks, rather than the first agenda item.

The ideal way to devise a reasonable lighting budget is to use standard, no-nonsense fixtures in all but a select few public areas such as lobbies, hallways, and conference rooms. (Restrooms are usually in this category and have quite sophisticated lighting that they can easily do without. However, a slightly overlighted restroom is a good alternative because it lends the impression of cleanliness.) The rest of the spaces should have lensed 2×4, three-lamp fixtures in areas with a lay in ceiling and 8-ft two-lamp industrial turrets in shop or maintenance spaces. The specialty areas can have parabolic fixtures, downlights, coves, or wall washers as circumstances warrant. The idea is to locate the special fixtures in only the most important, visible spaces, and of such a quantity that the incremental cost above the really basic fixtures is so small that the owner can see how inappropriate it would be to cut them from the budget.

This, then is the basic scheme. The fall-back position (i.e., no specialty lighting) is to be used in the event of a slashed budget only. Beyond this, there are some nice things that the owner should be asked to sanction such as parabolic fixtures in offices using a CRT much of the time, electronic ballasts, and multilevel switching in offices with daylighting available. These are all very nice features to have, and they are quite easy to change on the plans, too. Usually little more than editing a schedule is required. Any conscientious owner would like the added value, but in good faith the electrical designer should still keep them in a separate category, as "aesthetic and energy saving enhancements," that can be submitted to a line item veto. In essence, these can be sacrificed to value engineering, leaving the priority embellishments for the second round.

This may seem like an extraordinary amount of trouble, if not a duplicitous scheme, to go through to retain a few special lights in a relatively small percentage of the floor area. Yet, if you notice, the architect goes through the same elaborate process to preserve the most public, and visible, aspects of his or her work: the main entrance, lobby, and the important, focal points of the exterior facade. These are the crucial items that make the project unique and that market the architect's expertise to the public, the owners, and the staff. They are also what will become to the owner the small things that create a unique work environment and that add intrinsic value to the building and to the business. This translates into higher rents that can be charged to tenants and higher fees that can be charged to valued clients served by businesses housed in the building. So, in the end, the architect is preserving what is most important, long term, to the client and to the community.

The same philosophy applies to the electrical lighting in the highly visible public areas. In a sense, there is a double incentive to provide first-class lighting in these areas. The first is because they are the public spaces, and the second is because the architect has gone to considerable effort to embellish these areas, and quality lighting will bolster and highlight his or her efforts that much more.

The Lighting Palette

A fundamental selection of fixtures to be used on a sizable commercial lighting project is listed below in order of increasing impact and cost:

2×4 lensed

1×4 lensed

2×4 parabolic

Cove lighting with an eggcrate grill

2×2 parabolic

Pendants for indirect lighting

Recessed down lights

Wall washes with parabolic-shaped reflector

Wall sconces

This list of lighting options shows the relative cost of illuminating a space to the same light level using different types of common fixtures. Some fixtures may, per unit, be less expensive than fixtures ranked lower, such as cove lighting versus 2 × 4 parabolic fixtures. However, because the parabolic luminaire is much more efficient in its distribution of the light created, it takes fewer fixtures than using indirect cove lighting. Also, cove lighting and other fixtures installed in hard ceilings require more materials and labor than fixtures installed in lay-in ceilings

Generally speaking, the more elaborate fixtures are best used in specialty applications where they will be noticed the most. Wall sconces or recessed downlights used for illuminating a corridor are not cost effective or even aesthetically proper, unless the rest of the space has a very high quality of lighting as well. The same might be said for any space with a significant outside view—elaborate lighting is wasted in all but the few hours the space might be occupied when it is dark outside.

Daylighting Strategies

A delicate balance between window area and placement and space comfort is seldom accomplished. The trend is to maximize perimeter glazing, regardless. There are several detrimental aspects to this approach:

- On a bright day the illumination from the window can be 4 to 5 times that required by prevailing standards. This can cause eye strain, stress, and anxiety in the occupant(s).
- Venetian blinds are usually installed on large window areas, and a high proportion of them are pulled most of the time. This greatly reduces the elegance and effect of an unbroken window expanse. It also looks poor when the blinds get old, bent, dusty, and uneven.
- Often retail outlets with a large merchandise window wall must tint the glass so darkly that it is hard for passers by to see inside. It would be more effective in marketing wares to have a smaller glass area, with less tinting.
- Spaces with many windows are very hard to keep comfortable, summer or winter. Even though the space temperature may be at setpoint, the radiant energy from the window tempers the occupants, so they feel colder in the winter and warmer in the summer.

In the winter a tall window can cool the air right next to it, causing it to cascade steadily off the surface into the room. Baseboard heat is the only effective remedy to this.
- If there is not a pleasant view from the space, a large window can actually be detrimental to the overall effect of the office space.
- All things considered, large window areas are costly to construct and make spaces very costly to condition properly and consistently. Even small variations in temperature can have a marked effect on the attitude and productivity of the occupant. Thus, what a large window gains in visual appeal and implied status, it loses in the comfort and overall productivity of the occupant.

Light Shelf

An engineering and architectural challenge is to provide daylighting to as many spaces as possible. Alternative building and space configurations can bring the healthy, beneficial effects of natural light to many areas of a building other than just perimeter zones. Even a small amount of natural daylight can have important consequences in a workspace for ideal color rendition and visual comfort.

An atrium in the middle of a large open workspace is one effective way to introduce daylighting to a large area in the middle of a building. A glass wall around the atrium, floor to ceiling, creates a nice visual relief for the space. High ceilings in the surrounding large open office area provide for deep penetration of the daylight into all the surrounding spaces. Modular furniture with 5-ft partitions gives privacy but does not block the light from the work spaces most distant from the atrium.

Another approach is illustrated in Fig. 8.1. This technique uses high ceilings in outer offices, with a reflective light shelf to project more light to interior spaces. In this particular arrangement, where separate individual offices are desirable, there are windows above 7 ft so that light can be focused across an intervening hallway and into the interior space. The glass maintains privacy and quiet but permits the use of daylight through much of the day. The high office ceilings also endow the spaces with a sense of space and permit some novel lighting designs, such as pendant-hung fluorescent lights, to further enhance the visual environment.

Skylights are another relatively inexpensive source of daylighting, with striking visual affect. Architects prefer to use them in lobbies and main hallways, where they certainly look nice but do not contribute much to anybody's workspace. Nor do they save as much energy in hallways because the required light level is always much lower in

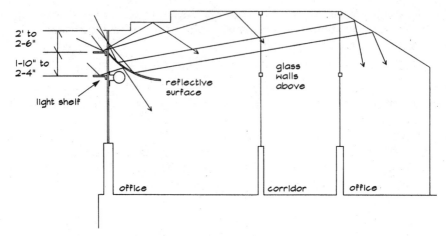

Figure 8.1 Daylighting concepts with light shelves.

corridors than in workspaces. It is better to use skylights in open office environments where individuals at work can benefit most from the natural light throughout the workday.

The benefits of natural daylighting are more than just reduced lighting costs and high-quality visibility. There is also a comforting connectedness to the outdoors and the slow and steady passage of time as the sun moves across the sky, the direction and intensity of illumination changing with it. These features give the space a higher quality and aesthetic value, with concomitant effect upon worker comfort and productivity.

The Nature of Light

One of the most important characteristics of light is that the intensity of illumination falls off as the square of the distance from the point source. As illustrated in Fig. 8.2, if the intensity is one unit at 1 ft from the source, it will be one-quarter of a unit at 2 ft and one-ninth of a unit at 3 ft, and so forth. As an application of this concept, consider a recessed downlight that provides 40 fc of light when installed in an 8-ft ceiling. If the same fixture is installed in an open lobby with a 16-ft ceiling, it will only provide 10 fc of light. This is the case for all types of light, from any source, although light focused with a fresnel lens or a parabolic diffuser is slightly less diffuse at the work surface. This is called the inverse squared law. Another interesting fact is that the light of the full moon is about 5 fc, whereas the brightness of the sun is about a million footcandles.

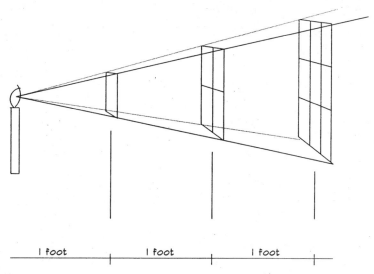

Figure 8.2 The inverse square law illustrated.

Zonal Cavity Calculations

The most common way to determine the average light level in a space is the zonal cavity method. The first step is to find the room cavity ratio (RCR), which is defined by the following equation:

$$\text{RCR} = \frac{5 \times H \times (L + W)}{L \times W} \tag{8.1}$$

where RCR = room cavity ratio
H = height of the room
L = length of the room
W = width of the room

This RCR is then used to extract the coefficient of utilization (CU) from reference tables for the fixture. The other values needed to find the CU are the reflectances of the walls, floor, and ceiling. These are available from the manufacturers of the surface covering or from a table such as Table 8.1. It is best to use the reflectance of the actual material used because there can be great variations between different manufacturers and products.

One very important thing to keep in mind when performing lighting calculations is the type, quantity, and color of furnishings that will be installed in a room. An office with three white walls and an exterior window wall will have a lower light level (at night when there

TABLE 8.1 Percentage of Light Reflected from Typical Walls and Ceilings

Surface	Class	Color	Reflectance (%)
Paint		White	81
Paint	Light	Ivory	79
Paint		Cream	74
Paint		Buff	63
Paint	Medium	Light Green	63
Paint		Light Gray	58
Paint		Tan	48
Paint		Dark Gray	26
Paint		Olive Green	17
Wood	Dark	Light Oak	32
Wood		Dark Oak	13
Wood		Mahogany	8
Cement		Natural	25
Brick		Red	13

is no daylighting from the widow) than the same size room with four white walls. If one or more walls are lined with bookcases or large wall coverings, this will diminish the ambient light level. The most dramatic effect occurs in large open spaces. After they are filled with modular furniture and cloth covered partitions, the ambient light level can drop by 50 percent or more.

The next variable in the zonal cavity function is the maintenance factor. This accounts for the diminished output from the bulb as it ages and is called the lamp lumen depreciation (LLD). Some representative figures for LLD are provided in Table 8.2. This value is further modified by another factor to account for dust on the lamp, the reflective surfaces, and dirty lens (if any).

The maintenance factor is the product of the LLD and an estimated dirt factor, in the range of 0.85 to 0.95. This will be higher for lensed fixtures and for ones with large reflective surfaces such as lensed and parabolic fluorescent fixtures, respectively.

The actual footcandles in a space is calculated by the following formula:

$$\text{FC} = \frac{L \times \text{CU} \times \text{MF}}{\text{area illuminated}} \qquad (8.2)$$

where FC = footcandles
L = total lumens output by all luminaires in the space
CU = coefficient of utilization
MF = maintenance factor (LLD × dirt factor)

TABLE 8.2 Lamp Lumen Depreciation (% of initial lumens at 70% of life)*

	Incandescent Lamps	
Lamp description	Power rating	LLD factor
General service	10 to 150 W	91
	250 to 500 W	90
	750 to 1500 W	86
Silver bowl	200 to 500 W	75
Reflector	R40	86
	R52 and R57	81
Projector	PAR 38 to 64	84

	Fluorescent Lamps		
	Hours per start		
Fluorescent	6	12	18
Instant start 425 mA			
Standard colors†	88	87	85
Improved color types‡	82	80	78
Rapid start 430 mA			
Standard colors†	87	86	85
Improved color types‡	81	80	79
Rapid start 800 mA			
Standard colors†	81	79	77
Rapid start 1500 mA			
Tubular	76	74	72
Others	70	68	64

*Even though these values are slightly dated, they give conservative figures for all the lamps on the market. Select a lamp that is closest to the target lamp, or less efficient, and use the given lumen depreciation factor in the footcandle calculations.
†Standard colors = cool white, warm white, white, daylight.
‡Improved colors = deluxe cool white and deluxe warm white.

Levels of illumination

There are many tables published by the Illuminating Engineering Society (IES) that describe the amount of light required in spaces of all sorts. There are also factors that specify higher light levels for older people and for performing critical tasks or those which are done on a low-contrast work surface. Otherwise, some general principles apply to the illumination of any space:

- The highest footcandle level is required for tasks that are very detailed, low in contrast, high in speed, prolonged in duration, or critical in nature. Beginning at 100 fc, the most tasking work can require 150 fc or more. For tasks of this nature it is best to research the most current literature to find the exact amount of light required for the specific application.

- In general office applications, where speed is not a critical factor, a minimum of 50 fc is recommended.
- Conventional industrial and commercial applications need 20 to 50 fc, with supplementary lighting as required for special tasks.
- General lighting systems are most often used for recreational activities and for familiar tasks of brief duration, with lighting provided on the order of 10 to 20 fc.
- In spaces where large objects are involved, with slow movement and good contrast, 5 to 10 fc is usually sufficient.
- Passageways with light traffic and no hazards need only 5 fc if provided in a uniform manner and of a good quality.

The following Tables 8.3, 8.4, and 8.5 provide additional information. In every case it is appropriate to provide a much lower ambient light level in the space via ceiling-mounted fixtures. The balance of illumination is then provided by task lights or by specialty lights when over 100 fc is desired (e.g., dental exam lights or operating room lights in a hospital).

TABLE 8.3 Approximate Reflection Factors

Medium value colors	Reflectance (%)
White	80–85
Light gray	45–70
Dark gray	20–25
Ivory white	70–80
Ivory	60–70
Pearl gray	70–75
Buff	40–70
Tan	30–50
Brown	20–40
Green	25–50
Olive	20–30
Azure blue	50–60
Sky blue	35–40
Pink	50–70
Cardinal red	20–25
Red	20–40

TABLE 8.4 Spacing and Mounting Height for Illumination Uniformity

Ceiling Height (ft)	Indirect—distance from walls	Semidirect—maximum spacing of luminaires	Light distribution (ft)					
			General diffusing—mounting height of luminaires	Direct-indirect—distance from walls	Spread direct—maximum spacing of luminaires	Semiconcentrating—maximum spacing of luminaires	Concentrating direct—maximum spacing of luminaires
8	2	9	8	2	7.5	5.5	2.5
9	2	10.5	9	2	9	6	3
10	2.5	12.5	10	2.5	10.5	7	4
11	2.5	13.5	11	2.5	12	8	4.5
12	3	15	12	3	13.5	9	5
13	4	17	13	4	15	10	5.5
14	5	19	14	5	16.5	11	6
15	5	20	15	5	18	12	6.5
16	6	22	16	6	20	13	7
18	6	24	18	6	22	15.5	8
20+	7	28	20+	7	25	17.5	9

TABLE 8.5 Practical Hanger Lengths for Suspended Luminaires

Ceiling height (ft)	Room width (ft)	Hanger length (in)	
		Offices and classrooms	Drafting rooms
7	7	—	—
	14	—	—
	28 and up	—	—
8	8	—	6
	16	—	6
	32 and up	—	6
9	9	6	12
	18	6	12
	36 and up	—	6
10	10	18	21
	20	12	18
	40 and up	6	12
11	11	21	21
	22	18	21
	44 and up	12	18
12	12	21	21 or 24
	24	21	21 or 24
	48 and up	21	21 or 24
13	13	21 or 24	24
	26	21 or 24	24
	52 and up	21 or 24	24
14	14	30	30
	28	24	24
	56 and up	24	24
15	15	36	36
	30	30	30
	60 and up	24	24
16	16	42	42
	32	36	36
	72 and up	30	30
18	18	42	42
	36	36	36
	72 and up	30	30
20	20	54	54
	40	42	42
	80 and up	36	36

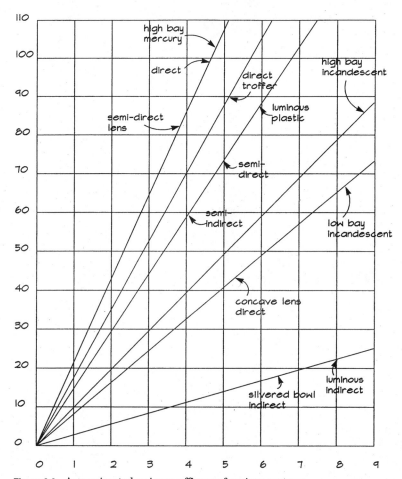

Figure 8.3 Approximate luminous efficacy of various systems.

Chapter 9

HVAC Equipment

The electrical systems for sizable commercial buildings have three main components: lighting, outlets, and heavy equipment. This chapter discusses the latter, the system that typically gets the least attention from the designer. This is disproportionate to its complexity, cost, and energy use. Once the electrical designer becomes more familiar with these implications, the equipment branches should get the engineering attention they deserve.

Before introducing the design issues, there are some politics of the production scheduling to consider. Most organizations have separate architectural, mechanical, and electrical groups. Coordination between them is such that often the hvac equipment schedule is not provided to the electrical designer until it is almost time to issue the construction documents. Many items are not specified until the very last minute, especially auxiliary equipment such as exhaust fans and ventilators. Sometimes not even the major air handlers, cooling towers, or chillers are finalized until the last possible moment.

There is really no reason for this delay. Just as the electrical designer selects the lights before beginning the lighting footcandle calculations and fixture circuiting, the hvac designer does the building load calculations. Based upon their peak load, the chillers, cooling towers, and condenser water pumps can be selected with confidence. Don't let the mechanical designer tell you otherwise. Besides, there are so many generalizations and safety factors in the loads analysis that many changes can be made to the floor plan, the occupancy, and the exterior exposure of the space without affecting the size of the equipment. If the designer tells you otherwise, he or she is not doing a very professional evaluation.

Most of the larger auxiliary equipment can also be selected with confidence. This category includes air handlers, ventilation fans, and mixing boxes. All of this can be done in the design development phase of the project, at about the same time the electrical designer is selecting fixture types and locating duplex power and computer/phone outlets. Then, while the hvac designer is laying out ductwork, piping, and diffusers, the electrical designer is circuiting the outlets, lights, and equipment.

There may be a problem if the hvac designer assumes the electrical designer needs the exact position, amps, and model number of the equipment. The electrical engineer, though, can do quite well without the exact nameplate data. Motor size and approximate equipment location is sufficient to get the basic equipment on the plans. It is usually not even a problem if the location shifts a few feet this way or that. Even changing the motor by a size does not usually affect wire, conduit, or disconnect sizes.

The two disciplines must recognize their respective priorities. The electrical designer has fully twice the volume of plan work (lighting, power, signals, and equipment sheets) as the mechanical designer (hvac and sometimes a piping plan). Exact locations of items are not as important to the electrical plans as their physical inclusion and schematic circuiting. A small branch duct or a few balancing dampers can be left off the hvac plans without incident because the flaw is usually covered in the specifications. A missing motor is another matter altogether. The cost of the motor, disconnect, feeder, and circuit breaker—even to the point of upsizing the panel and its feeder—can be substantial. A missing wire size or disconnect size is much less of a problem, as long as the apparatus is generally drawn up on the plans or the riser. Electrical designers cannot delegate any authority to their specifications, either, as hvac engineers can. If it is not on the plans, it is not important, in most electrical matters. All of this must be made quite clear to the hvac staff so that even the most rudimentary information is provided to their electrical counterparts fairly early in the production process.

The electrical engineer should strive to have a majority of the design down on paper and organized on the riser, in the circuiting to distribution panels, and the organization of wireways or busways long before the final documents are to be issued. This leaves time in the final frantic days or hours for checking the plans versus the final mechanical plans and getting power to those few items that inevitably must be specified at the last moment.

The next sections will detail some important design guidelines for delivering power to some of the most common hvac systems. There are a few principles that apply in all instances:

- Use the highest voltage available. This reduces the size and cost of wire, conduit, disconnects, and circuit breakers.

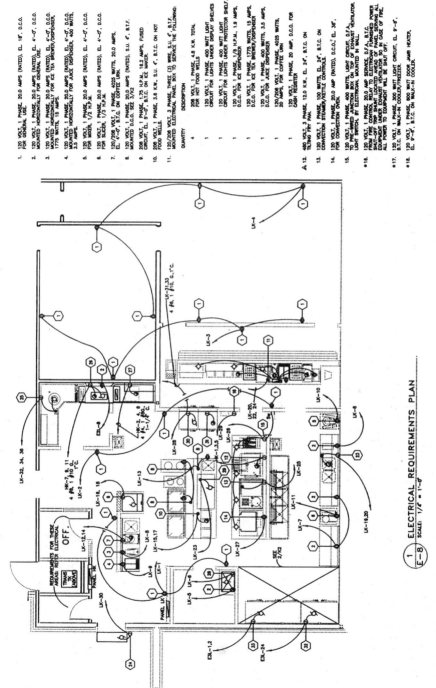

Figure 9.1 Kitchen layout.

GENERAL NOTES

1. ALL OUTLETS SHOWN TO BE W.P., G.F.I., MTD 15" AFF
2. POWER CIRCUITS FOR OUTLETS SHOWN THIS PAGE NUMBERED ON SHEET E-3
3. SEE SHEET E-1 FOR SYMBOLS.

KEYED NOTES

① UTILITY COMPANY MAIN SERVICE TRANSFORMER
② EMERGENCY GENERATOR, SEE DETAIL.
③ EMERGENCY GENERATOR AIR DISCHARGE.
④ 6" HIGH STRUCTURAL CONCRETE SLAB. CLASS "A" TYPE CONCRETE, 3000 PSI AT 25 DAYS, WITH #3 REBAR 10" ON CENTER, 1" FROM TOP SIDE & END OF CONCRETE.
⑤ MEDICAL GAS STORAGE AREA
⑥ MEDICAL GAS ALARMS
⑦ WALL MOUNTED LIGHT, HUBBELL MODEL NUMBER PRS-01004-118, WITH PBT-1 PHOTO CONTROL, PVL-PK GASKET/WIRING KIT AND 100 WATT METAL HALIDE LAMP. FIELD ADJUST FIXTURE TO PROVIDE THE BEST LIGHT TO THE AREA AROUND GENERATOR. FIXTURE TO BE MOUNTED FLUSH WITH TOP OF WALL.
⑧ INSTALL A 3" EMPTY CONDUIT FROM JUNCTION BOX ABOVE CEILING TO PANEL MDP. PROVIDE AND INSTALL A PULL LINE IN CONDUIT. LABEL BOTH ENDS. JUNCTION BOX TO BE 24"x24"x8" DEEP, NEMA1-ONE SCREW COVER.

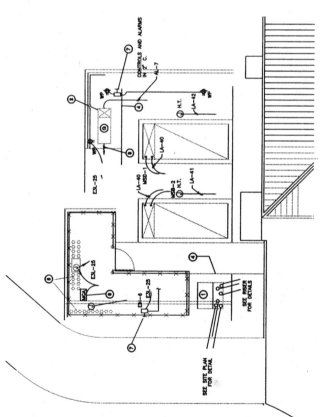

Figure 9.2 Utility yard layout.

HVAC Equipment   123

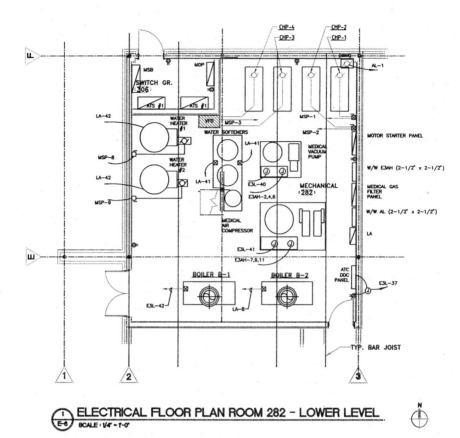

Figure 9.3 Mechanical room layout.

- Use three-phase equipment when it is available in the size specified by the hvac designer. This equipment is often more efficient and dependable.
- Where several items are located in a space, instead of running a feeder to each from a remote panel, combine the service. Route a single feeder to a main disconnect and then a wireway (or small panel with a main breaker, if it can be justified). Then bring home runs from each piece of equipment to this location.
- Specify a motor control center if there are more than three large motors in a single space.
- In the interest of energy economy, the hvac designer should have specified direct expansion (DX) equipment that is either a straight heat pump or one with a gas furnace. The only instance where electric resistance heat should be specified is on heat pump units. It is

needed then as an auxiliary for use only in very cold weather and if the unit is too small for a unit warm-up cycle. When installed, the electric resistance heat should be installed on a breaker separate from the unit itself so the breaker can be locked open during the cooling season to prevent reheat.

The electrical designer should educate the hvac designers about these guidelines so that the best equipment can be selected on each project. Conversely, the electrical designer should know enough about interpreting hvac equipment cut sheets to discern the voltages and phases available. Then specific demands can be made of the hvac trades to provide the necessary equipment.

Packaged Rooftop Equipment

The easiest, cheapest, and most practical hvac design for a small commercial system is the packaged rooftop system. This requires a single-point power connection at each unit. There is a preferred way to do even these simple installations. Some of the better design characteristics include

- Do not route any conduit over the roof. It's messy and causes problems later when the building is reroofed. It is better to keep all feeders below the roof in the ceiling space, feeding up to the unit through the unit curb.

- Provide a remote disconnect for the rooftop unit. If the disconnect is mounted on the unit itself, it is a problem removing the unit for maintenance or doing any work on the power service. The disconnect can either be mounted on a nearby wall or on a sturdy unistrut rack. This disconnecting means must be within line of sight and within 50 ft of the unit.

- Often penetrations must be made through the roof to bring a feeder up that is remote from the unit at the disconnect location. It is good practice to use a standard sheet metal curb rather than a pitchpan. A Thycurb or something equal is far less leak-prone and more dependable, for the modest extra cost.

- If there are several rooftop units in the same general location, it is more economical to install a small NEMA 3R (waterproof) panel in the approximate center. The circuit breakers—as long as the panel is in sight and within 50 ft of all units—can serve as the unit disconnects as well as the circuit protection. Again, the panel should be on a sturdy unistrut rack.

- It is always a good practice to provide a weatherproof/ground fault isolation (WP/GFI) outlet at each unit location. If it is a very

large unit, or several units are in the same vicinity, a switched light is also helpful. These are code requirements in some cities, and they are always very helpful for safety as well as maintenance purposes.
- The last few feet of connection to all mechanical equipment should be flexible, waterproof metal-clad electric cable. This is for easy maintenance and prevents any mechanical vibrations from being transmitted to the conduit from the unit.

Split Systems

The next most common arrangement is a small mechanical room with several vertical, floor-mounted air handlers serving dedicated zones in the building. This is a favored arrangement in smaller facilities because the maintenance is easier than a central system with variable air volume (VAV) boxes or air handlers and a chiller.

When such systems are installed in a school, library, or other building with a high occupancy, a large quantity of outside air must be delivered to each space by the air handlers. This is the one shortcoming of small split DX systems: They cannot handle more than about 25 percent outside air, and even that volume is pushing the limits of the equipment.

An economical solution is to use a single unit to pretreat the outside air for all the air handlers in the mechanical room and then duct it to each unit. This special unit must have two cooling coils and two condensing units. Electrically, it must also have a two-stage thermostat to sequence the condensing units. Sometimes it must even have a reheat coil to raise the temperature of the outside air after it has been dehumidified.

Another common feature is to control all the air handlers by a single time clock, with a 0- to 3-h override timer next to each unit's remote thermostat. This allows use of the system outside of the scheduled hours. Finally, the central time clock can be overridden by a remote "freeze stat," which is a remote temperature-sensing bulb located outside. When the outside temperature drops below 38° or so, all the units are automatically cycled on. This keeps plumbing and other items in the facility from freezing.

These controls details are provided here because on smaller jobs they are often the electrical contractor's job to install. The work is too small to warrant a separate controls contractor. Also, because the electrician is already on site to do all of the main wiring, he or she is in a position to bid the controls work much more cheaply than a controls contractor. So it helps to become well versed in the simpler, more common controls schemes for smaller jobs.

The power supplied to the hvac equipment in this instance should comply with several requirements:

- Instead of a disconnect at each air handler, install a small panel in the room. The circuit breakers can double as the disconnecting means if the panel is within 50 ft and in sight of all the equipment.
- This same philosophy applies to the outside condensing units. The disconnecting means should not be integral to the equipment but should be on a separate unistrut rack or the breaker in a remote NEMA 3R panel.
- It is always a good idea to be sure even small mechanical rooms with many items of equipment, disconnects, and panels are ventilated. The best way to extract heat from the space is with a thermostat-controlled ventilator on an exterior wall, drawing replacement air in through door louvers. The hvac designer usually provides such apparatus in large mechanical rooms but rarely in small ones even if there are several vertical units. If the electrical designer does not assert this need, the heat trapped in the room can reduce the operating efficiency of the equipment and even its useful lifetime.

A variation on this theme can occur when several small units are used to air condition a single large space. Some common examples are gymnasiums, cafeterias, and libraries. The controls can be demanding. For example, consider a gymnasium with two air handlers and a single two-stage thermostat. The sequence of operation for minimum energy use and optimum space conditioning is

1. On a call for cooling both air handler fans cycle on, but only the compressor(s) of one unit are on line. All fans are turned on to ensure full movement of air through the entire space.
2. If the load cannot be met, the second air handler compressor cycles on until setpoint is achieved. This can be done by a single two-stage thermostat, and the system must be configured so that the off-line unit does not cycle on in heating load as the other reduces space temperature.
3. The wiring should be installed so that the lead unit can be switched, thus preventing excessive wear on one of the systems.

Air Handler Rooms

In a facility with a central chiller, individual air handlers are situated around the building in special rooms. Hot and cold water is piped to these air handlers, where it is used to temper air delivered to the adjacent occupied spaces. Older units have a single-speed fan, with damper-actuated inlet vanes to modulate the flow air through the unit. Most large new systems have a variable-speed fan, if it is a variable-volume system. This is a dependable, efficient system that is popular even for smaller installations.

Often electrical distribution equipment is located in these remote air handler rooms. Because they are evenly distributed throughout the floor plan, air handler rooms are well situated for the electrical system, too. If this is the case, some careful coordination with the hvac designer is necessary. There must be at least 3 ft of clearance in front of panels, which must be maintained even when the air handler is undergoing maintenance. This includes pulling the coils out of the unit for inspection or renovation, replacing the fan motor, and changing out filters. There should also be no ductwork above or near the panel location(s). This allows room for conduit runs to turn straight down into the panel. No hydronic piping should be routed above the panels either.

Variable-Frequency Drives

Many of the large pieces of mechanical equipment are being installed with variable-frequency drives (VFDs) to match motor speed to the required load or air flow rate. The reason for this is that the savings in energy use can be substantial. The horsepower is a function of the cube of the air flow rate. This means a 10 percent reduction in air flow delivered by a fan requires 0.9^3 as much energy, or about 75 percent as much as the fan at full speed. The air handlers are designed for maximum cooling (heating) load, so the actual demand for air falls far short of 100 percent most of the time. Even a small speed reduction in a large horsepower fan that operates for many hours a year can add up to great savings quickly. The extra cost of a VFD can be paid off very quickly under such circumstances.

The one drawback of VFD systems is that when the device is out of order or extracted from the circuit for maintenance or replacement, the air handler fan motor is out of order, too. This is always a problem for building owners. So the circuit should be designed with a manual bypass that will keep the motor running at full speed when the VFD is inoperable. The air handler controls will have to receive a signal to this effect. Otherwise the hvac system will not operate at full speed. It's much better to have more air than none at all.

The best way to circuit such a system is to have a disconnect as well as a VFD for each large motor. The power is fed from the nearest panel to the disconnect and then to the VFD that feeds the motor. The disconnect is therefore readily available to do work on either the VFD or the motor. Also, if the VFD is removed and the motor is running full speed on the bypass, the disconnect still provides full protection to the circuit as required by code.

Central Plant

In most large commercial installations, the central plant is the single greatest power demand. This is where the chiller, pumps, and boilers are situated. There is almost always enough equipment to justify a motor control center (MCC) for all the starters and/or VFDs. These are large, expensive items, so it is important to not only size them carefully to match the existing power requirements but also to anticipate any future needs. These often stipulate a higher total rating, with larger bus bars and grounding capacity, plus a few extra spaces in the MCC itself.

Although it can lead to sole sourcing the device, it is good design practice to build the VFDs into the MCC. This makes for a neater, simpler installation that is more accessible for maintenance and repairs. Also specifying the capability for metering the voltage and amp draw of the entire unit, or individual components, is a valuable resource to facility operators. Many new systems allow this information to be ported directly into the building energy management system for continuous monitoring of these important functions.

The benefits of monitoring individual pieces of equipment, which can also be specified as a direct readout from most electronic devices such as VFDs, far outweigh the small additional initial expense. The facility engineer can develop a rigorous profile of energy use by all the major items of equipment, which is important for diagnostics and troubleshooting. They also provide important data for energy conservation retrofits and for capacity analyses. For example, pump head and flow changes with each modification to the hydronic system (i.e., adding a new wing or additional cooling load), and it is important to have some ready means to track the performance of the reconfigured system.

Often the construction budget will not pay for such small, if useful and ultimately cost saving, enhancements to equipment. Even so, the equipment that is installed should be designed for future provision of such monitoring needs. For example, it does not cost any more to provide a communications port on the VFD. Even if the device is not hard wired into a monitoring loop upon installation, the port can be used in the future for either local monitoring with a portable meter or wired into a building energy management harness.

There is one final aspect of VFDs that causes a lot of confusion when jobs are bid and built. The specifications should clearly state which discipline is responsible for providing, installing, and wiring the VFDs. It is best for the mechanical equipment manufacturer to provide the VFD itself because it must be carefully matched to the motor. The electrical contractor is the best one to wire power to the whole system, with the exception of the controls, which should be done by the controls contractor. This division of labor has each discipline doing what they are best at and should minimize the necessary coordination in the field. This does not, however, relieve the electrical designer of the responsibility of coordinating with the hvac designer to ensure that the proper VFD and all the necessary accessories are specified for the application.

Cooling Tower

A large installation can have hundreds of horsepower of motors driving the fans of the cooling tower cells. Often, in older installations, the fans are staged according to the cooling load, but they run all on or all off. A fairly recent modification of this is to have two-speed fans. This saves a lot of energy, but it can wear the bearings and motors considerably if there are frequent start-stops. An alternate system (which costs more to install but also saves a little more energy in operating costs) is to have one fan motor driven by a VFD.

The electrical engineer is usually not consulted by the hvac designer about choice of motors and controls for this equipment. So it is up to the electrical designer to ensure that high-efficiency motors and adequate controls are specified. This is particularly important when an existing single-speed motor on a cooling tower fan is retrofit with a VFD. The drive should be carefully selected so that it matches the motor characteristics closely. If this is not possible, the motor should be replaced with one that is designed to be operated by a VFD.

There are a few more small electrical needs for the cooling tower installation. Most cooling towers also have exposed piping that must be heat traced. A single 20-A circuit will suffice in most cases. Also, some towers have an electric basin heater, dampers, and other controls that require power. These needs should be researched carefully with the help of the hvac designer and the manufacturer. Finally, it is always helpful to have a WP/GFI outlet near the tower for use by maintenance personnel.

Miscellaneous Equipment

There are many mechanical items that require power: fans, motors, controls, and pumps. These items are usually integral to hvac equipment, but they are available in a range of sizes, capacities, and configurations. Although the mechanical designer is able to size the equipment so that it performs its job properly, the electrical items are not always the best for the application. It is important for the electrical designer to investigate this equipment and to offer constructive suggestions to the hvac designer so that the total system can be ideal in cost as well as ease of maintenance and operating expenses.

A good practice is to clearly designate each item of equipment on the plans. Using keyed numbers is an economical method. Even if a specific product is not identified, such as a junction box installed for a garbage disposal or garage door opener, if the information is available and was used in the design of the power source, it belongs on the plans. This can avoid a lot of confusion later on if the owner changes the location of the disposal or garage door by change order; it will be simple enough to do the same with the electrical service.

VAV boxes

Mixing boxes are a common mechanical item that is often left off the electrical circuiting plans. Even if the devices are not fan powered, they still require some electricity if the damper actuator is electric (rather than pneumatic). Most new installations have electric controls so that the devices can be monitored and controlled by the building's energy management system. Even if there is no such system at the present, it is prudent to allow the flexibility to install one in the future.

Some smaller systems have electric reheat coils in each VAV box. It is generally preferable to install hot water coils in the boxes if the hot water can be generated from gas or fuel oil. This is a much more economical heat source. Otherwise, there are a few things to keep in mind when circuiting coils in VAV boxes:

- Circuit the electric resistance heaters on separate breakers so they can be turned off in the summer as an energy conservation measure. Areas with low-humidity requirements or moisture control should be on separate breakers from the other spaces. These reheat coils may be needed year round to control space humidity, whereas the rest of the coils in the building can be turned off. The building owner may even choose to keep them off during the winter or to limit their use to extremely cold days. Using the central heat source is always more economical. If the building was zoned carefully by the hvac designer, the VAV reheat coils may rarely be used anyway.
- Because the coils can be fairly large in some boxes, a safe design practice is to provide a motor rated toggle switch as a disconnect at each box.
- The control power for several VAV boxes can be circuited together. The boxes may be far apart in some installations, so it is important

to be wary of voltage drop through the line. The boxes at the far end of the run may not have enough voltage to operate properly, and this will cause anomalies in the space temperature that the controls contractor may not be able to diagnose.

In installations where the hvac equipment is on emergency power, the controls as well as the power to the VAV boxes should be on the emergency power, too. This includes the VAV box control power as well as the power to the reheat coils.

Boilers

Gas or fuel oil boilers usually have a single-point electrical connection. In some instances, however, a small circulating pump may need to be interlocked with the boiler controls via a flow switch or other sensing apparatus. This keeps a flow of water through the device, preventing the heating elements from operating if there is no positive flow through the tubes.

Electric boilers should not be used for high-volume loads. If fuel is not available, it is more economical to install a storage vessel on site than to provide the feeders, panels, and resistance heating elements for the boiler. The operating costs are certainly less, unless the facility has a very short heating season or minimal hot water needs.

Either type of system should be installed with easy-to-access thermostat controls. When the weather is warm, the hot water temperature can be increased. Large facilities can justify an automatic system, with a remote thermostat to modulate the water temperature. Otherwise, the maintenance staff can do so manually as the seasons change each year.

Roof ventilators

An attic space in a small commercial building that is force ventilated can add an R-value of 5 or more to the effective thermal resistance of the roof assembly. The reason for this is that an attic that is only naturally ventilated can have temperatures 30 to 50° above ambient in the summer. Of course, this cooling is obtained at the cost of operating the fan. Consequently, the fan is often a variable-speed fan controlled by a thermostat in the attic space. This arrangement runs the fan only when necessary and at the slowest speeds possible to ventilate the space.

A common application is to interlock the attic fan with the building air handler so the fan runs whenever the building is occupied. This simple arrangement is done by the mechanical designer to reduce the R-value of the building, which means smaller hvac equipment and lower cooling operating costs. No account is made of the power used by the fan or how much this energy use can be lessened by some very simple controls. It is contingent upon the electrical designer to ensure that proper consideration is given to minimize the long-term energy use.

Exhaust fans

Mechanical designers like to interlock individual restroom exhaust fans with the light switch for the space to save fan energy. Whenever the lights are on, the exhaust fan runs constantly. Depending on the use of the facility, however, this may not provide adequate ventilation, especially if the lights are on a motion sensor control. A better solution in either case is to have a 5- or 10-min delay on the fan after the lights cycle off, by whatever means.

In this instance, the fractional horsepower fan uses so little energy that turning it on and off frequently may not do much good. It not only wears the device out quickly but also uses almost as much electricity because of the high inrush current to get the motor up and running. This is a situation where the controls may be more trouble than they are worth, especially if it is a busy public restroom. It may, in fact, need constant ventilation to remain odor free.

A common use for restroom exhaust fans is to provide air conditioning in the space by drawing return air through the room. No air is supplied directly to the space. If this is the case, the restroom fan should run constantly. The electrical designer should be aware of this possible application before any sort of controls are installed on the fan.

Large buildings are often ventilated using a single fan for several restrooms. In this case the fan must be interlocked with the building time clock or central air handler so that the fan operates whenever the facility is occupied. This is the most economical use of fan horsepower and provides the most dependable ventilation of the restrooms.

Pumps

Powering up pumps is not a complex operation. When several pumps are staged for flow and dynamic head capacity, it can be a problem. Mechanical designers like to have two pumps piped in parallel to generate hydraulic flow for a system. One pump handles low-demand situations, and the second pump is brought on-line as the flow or head requirement increases. To save energy, one of the pumps is put on a VFD or two-speed motor. This gives some flow modulation when one pump is on-line, but after that both pumps must run at full speed anyway. The only way to accomplish modulation for the full range of flow is to install a VFD on both pumps. One VFD on two pumps does not work well either. So, no matter what the mechanical designer specifies, the only practical way to modulate several pumps is to have one VFD on each pump.

The situation gets even more complicated because pumps have unique efficiency curves. Several smaller pumps on VFDs, if properly staged, can use much less energy than a single large pump on a VFD. The small pumps must be staged on-line to keep them operating at peak efficiency, usually according to a flow sensor. This is called water-to-water efficiency. It is a complex operation that must be carefully designed and it should be outsourced to a specialist.

The pumps, VFDs, controls, and flow sensors are all predesigned and built on a skid unit, with a single electrical connection and simple piping connections. If the mechanical designer is custom designing the pumping system for a large application, this is the way to get the most effective design, the best controls, and the lowest first cost and operating cost.

Water heaters

There are two common ways to provide domestic hot water at a facility (increasingly, the same applies to some hydronic systems). The traditional way is to have hot water circulated constantly in the building loop so that each hot water tap is only a few feet from the loop. This means the user will only have to run a little water from the sink before hot water starts coming out. This system has a large circulating pump that operates all the time, as does the boiler in order to make up for heat lost through the pipe insulation as it circulates through the building. (Some facilities control the circulating pump with a thermostat in the water line, so the pump runs only when the line temperature drops below setpoint.)

A new hot water system is to heat trace all the hot water piping. Heat tape is a composite plastic material with 12-V wiring embedded. The strip is epoxy welded to the pipe beneath the insulation. The low-voltage power to the tape heats the water, using only enough current to achieve setpoint temperature. The heat tape is used on all the piping, so hot water is available at the tap without wasting any water waiting for the hot water to start.

There are several important advantages to this system:

- The piping installed is half that needed for a closed-loop circulating system. This greatly reduces initial cost.
- Because there is only half as much pipe, the standby losses and the heat lost from the circulating fluid through the insulation are also reduced by half or more.

- The system uses line pressure to deliver hot water, so there is no circulating pump. This saves the operating cost of running the pump constantly.

All told, the heat-traced system is less expensive to install and costs substantially less to operate than the traditional closed loop piping.

Variable frequency drives

A lot has been said in other sections about the application of VFDs. They can be used on almost any large motor at a facility: pumps, fans, chillers, and air handlers. In every case energy will be saved. However, the extra cost of the VFD is not always worth the savings in energy use. A good objective is a payback (cost divided by energy savings per year) of 7 years or less.

The next important considerations are the ability of the facility staff to maintain the equipment and the availability of parts. Even large facilities, especially in the public domain, shy away from VFDs because they simply do not have the expertise to keep them up properly. Inadequate maintenance can cause energy use to go up, instead of down. So it is important for the electrical designer to be familiar with the expertise of the facility staff and their comfort level with the sophisticated electronics of VFDs.

There are other alternatives to VFDs. Two-speed motors are popular and in most applications save almost as much energy as a motor driven by a VFD. The reason for this is that at low speeds the VFD can be less efficient than a motor running on low speed from fixed windings within the device. Another alternative, in pumping applications, is to stage several different-size pumps on line according to the demand. This operates several sizes of motors, each at its full speed and capacity, and is fairly easy to implement with common control devices and digital protocol.

Another important consideration is that when a prime mover for a system is on a VFD, it means there are many other variable devices in the entire loop. In an air handler, this means many VAV boxes throughout the building. A pumping system will have three-way valves everywhere. Even though the overall system will save on electricity, this is accomplished at a high price that includes many other control devices than the VFD. Such a complex system must be very well maintained by qualified staff, or it can potentially use more energy than the basic constant flow arrangement.

These considerations are often handled by the mechanical designer, with little consultation with the electrical engineer other than to notify him or her of the equipment needs. However, the facility staff may be quite capable of maintaining the VAV boxes or the three-way valves but unable to be as vigilant for the VFDs. If such is the case, it is up to the electrical designer to bring this to the attention of management so that the client is not left with an overly complex system that cannot be kept in good working order.

Controls

Although most hvac control systems are low voltage, with power distributed as part of the wiring harness designed and installed by the hvac controls contractor, there are a few items that require main electrical power. Usually each control panel has a line voltage input, with transformers inside to step down the power to the control voltage. Dampers on air handler inlet vanes, outside air louvers, and many other applications require line voltage. Many of these items are not even shown on the mechanical drawings, either; they are only shown schematically in a line diagram or as a performance requirement in the specifications section.

The electrical designer should identify any such instances and make a clear distinction in the project documents who is responsible for providing the power. A good practice is to assign to the electrical contractor all those circuits that are shown and circuited on the drawings. These are usually the larger actuators, motors, and control elements. All other electrical items required for the hvac controls are in the scope of the controls contractor. This should be clearly stated in the specifications, with actual elements indicated to the extent possible.

Integrated Systems

There is a trend to combine several hvac functions. The heat that is extracted from a building during the summer is transferred into water for use in the domestic hot water circuit. This makes both systems operate more efficiently. It also keeps the compressor operating at a lower temperature, so the maintenance and useful lifetime of the equipment is greatly extended. The challenge to the electrical designer is to not only understand these new systems but to provide input to the design team concerning the impact on cost and losses from the electrical power distribution system serving these devices.

Heat pumps

The technology to combine a home air conditioning and water heating system has been in use for many years, and the application is increasingly suitable for commercial use. The heat pump and water heater are combined so that during the cooling season heat is transferred from the occupied space into the domestic water heater. Because the air conditioning unit is a heat pump, this process can continue into the heating season. When the weather is very cold out, however, an electric or gas auxiliary heat source is needed for supplementary heating of the water.

The water side has a storage tank and a small thermostat-controlled circulating pump. Through interlocking controls, the pump operates when the heat pump is on and there is a call to heat the water in the tank. The airside is an exterior compressor/heat pump, with a refrigerant line going to the air handler inside. This refrigerant circuit also passes through a refrigerant-to-water heat exchanger, and heat is extracted into the water on the way outside to the compressor. Because the load on the compressor is reduced, the fan does not need to operate as often, and in some cases the compressor can be downsized.

The impact on the electrical system, for an all-electric facility, is a smaller feeder to the compressor and no feeder at all to the water heater (unless a small supplementary heat source is required in very cold climates). The annual electric bill is reduced because the compressor is smaller and the electric water heater is replaced with a small fractional horsepower circulating pump.

Desuperheaters

The system just described can be retrofitted to an existing heat pump installation by the addition of a desuperheater in the refrigerant circuit. This is a stock item that is predesigned by some manufacturers to specific hvac units. It is slightly less efficient than a factory-installed system, but it can have the same positive benefits for a hvac system that has many useful years remaining in its operating life.

Typically, this is done most economically at all-electric facilities that have a large hot water demand. Commercial kitchens, laundries, gymnasiums, dormitories, and hospitals can benefit the most from this inexpensive hot water source. Often there is a particularly large demand at some time during the business day that peaks the electric demand. Having an alternate heat source for domestic hot water will greatly reduce this peak and translate into significant savings on the utility bills.

Air-cooled chillers

This is another application well suited for commercial and institutional applications, such as hospitals, prisons, and military barracks. The latter, in particular, have a very large hot water demand once or twice a day. This gives the system up to 12 h to heat up the water in the hot water tank, often during the hottest hours of the day when the most heat is extracted from the building. In some instances the facility use can be scheduled to take advantage of this phenomena.

Large air-cooled chillers are well suited to the installation of an appropriately sized desuperheater. Only a single manifold is required, so the controls are not too complex. The advantage is that the fans do not have to operate as much if heat is extracted from the refrigerant into the hot water circulating system. This reduces the peak cooling load and allows the compressors to operate more efficiently. This effect further reduces the peak electrical load, as well as the standard demand usage.

Water source heat pumps

The traditional air conditioning compressor is a device that extracts heat from the refrigerant loop, discharging it into the atmosphere at the fan blowing air across the coils. A water source heat pump (WSHP) system extracts heat from the refrigerant loop by passing it through a heat exchanger with refrigerant on one side and water on the other. This is a more efficient heat transfer process than liquid refrigerant to air. Also, evaporative cooling at a cooling tower, where heat is extracted from the circulating fluid, is a more efficient process than cooling by hot outside air blowing across finned cooling coils.

Typically a large building has a single circulating loop from the cooling tower to the air handlers. A special unit can also extract heat from the same loop to create hot water. These devices are provided by the same WSHP manufacturer and are exactly the same as an air handler but with no fan. This unit takes heat out of the loop before it reaches the cooling tower, so this diminishes the load on the tower; therefore, the fans and pumps do not run as much.

A further advantage of the WSHP system is that it can operate on a boiler in the winter. If the facility is a large one with many interior zones, there is a need for cooling all year. The WSHP system will simply have the interior zone air handlers discharging heat to the loop and the exterior ones extracting it. This effectively transfers heat from one part of the building to another by virtue of pump and fan horsepower alone.

A final option available with WSHP systems is to install several ground loops in the system. These are either vertical wells or horizontal ones in which PVC piping is buried. The water circulates through this piping and discharges heat to the ground instead of at a cooling tower. The opposite happens in the winter, when the ground serves as the heat source instead of the heat sink. So, the installation does not need either a cooling tower or a boiler.

In most instances, however, it is better to size the ground loop to handle only part of the load. In southern climates the ground is best used to handle the heating load and a part of the cooling load in the summer. In northern climates that have a predominant heating load each year, the ground loop is sized to satisfy a part of the peak cooling load to reduce peak demand charges. In either case, these hybrid systems do more than carry a part of the hvac load all year; they also add a thermal storage element to the system, the ground acting as a large and flexible heat sink/source.

Again, the WSHP system reduces peak electric demand charges and usage charges as well. It is suitable for commercial use, especially in schools and colleges, which have a high interior cooling load all year.

Individual Control Elements

Each hvac system has many components in common. Every system needs a way to sense air and fluid temperatures. This data is then used to control fans, compressors, dampers, and valves. In days past, these actuators, sensors, and modulating controls were pneumatically controlled. Many new systems still use pneumatic controls, which use pressurized air as the controlling medium and force. Pneumatic controls are sturdy and dependable and are well suited for small jobs or at remote locations. However, electronic controls are becoming much more prevalent because they are more versatile and, increasingly, less expensive and more dependable.

It is most common for all the control items to be designed and installed by the hvac expert. However, there are frequent instances when a single damper controller or control valve is needed on a small project. The electrical designer who has the versatility to assist in this small area can save time and money on the project. The electrical contractor who is able to do so and who saves the expense of hiring a controls contractor is just as well thought of. There will be some electrical work to do on the job already, and so a low bid price on the controls portion will increase the possibility of being awarded the entire project.

Time clocks

Building owners are well aware that occupants are remiss in turning the hvac system off at night and on the weekends. So, they install systems to cycle the equipment off automatically outside of the regular business hours. There are many ways to do this. A single unit can be most economically controlled by a single programmable thermostat with an override timer. This device can be fitted with a locked cover or even a remote sensing bulb in the occupied space so that the thermostat itself cannot be tampered with by occupants. A separate override switch such as a 0- to 3-h device can be located in an accessible location.

Control valves

This common modulating system has three components: the motor, the controller, and the valve. There are many means by which force is transferred to the valve actuator. Smaller systems that do not require very accurate control use direct action. The rotating shaft of the motor causes the valve shaft to rotate, moving the valve up and down by rotating it in its housing just as the device is closed by hand. This method cannot develop a lot of torque, but it is inexpensive and has few moving parts to adjust or cause operational problems.

Most hvac applications need a lot of torque, or turning force, to rotate the valve under the pressure of fluid flow in a pipe. They also need to position the valve very accurately in order to control flow precisely. This combination of needs is best met by a mechanical device that moves the valve only slightly for many revolutions of the motor. This means a smaller motor can develop a higher effective torque on the valve shaft, and the motion of the valve can be limited to small increments sufficient to provide the desired control level.

There are as many ways to arrange this action as there are controls manufacturers. In general the fewer moving parts, the less maintenance problems the device will have. If there is a lot of vibration, any adjustments will surely rattle loose eventually. Also, if the valve is modulated frequently, and under high forces, the controller and the motor should be of heavy-duty components. Cam driven devices instead of long armed linkages are especially effective for such rugged applications. The cam will not go out of adjustment, whereas the linkages are sure to. If the environment is very dirty or caustic, the entire mechanism should be enclosed in a solid, sealed enclosure.

Grounding Conductors

Special attention should be devoted to properly grounding hvac equipment and any other large motors at a facility. Table 9.1 is the official NEC sizing table.

Neutral conductors, from the neutral connection of a transformer or generator, must be fully insulated and sized to match the current rating of the grounding impedance (and no less than #8 copper wire). The impedance is recommended at a value sufficient to limit the ground fault current to the capacitive charging current of the system.

Table 9.2 is a grounding table for hvac equipment. The first criteria for sizing a grounding conductor is to locate it as near as possible to the grounding connection. It should be sized to match the ampere rating of the overcurrent protecting device of the equipment. Next, it is important to take into account the length of the ground cable and its voltage drop. Long runs should be oversized to be sure the grounding

TABLE 9.1 Minimum Size of Equipment-Grounding Conductors

Rating of setting of automatic overcurrent device in circuit ahead of equipment; not exceeding (A)...	Size of conductor	
	Copper	Aluminum
15	14	12
20	12	10
30	10	8
40	10	8
60	10	8
100	8	6
200	6	4
300	4	2
400	3	1
500	2	1/0
600	1	2/0
800	1/0	3/0
1000	2/0	4/0
1200	3/0	250 MCM
1600	4/0	350 MCM
2000	250 MCM	400 MCM
2500	350 MCM	600 MCM
3000	400 MCM	600 MCM
4000	500 MCM	800 MCM
5000	700 MCM	1200 MCM
6000	800 MCM	1200 MCM

SOURCE: Reprinted with permission from NFPA 70-1993, National Electrical Code.

TABLE 9.2 Grounding Electrode Conductor for AC Systems

Size of largest service-entrance conductor		Size of grounding electrode conductor	
Copper	Aluminum	Copper	Aluminum
2 or smaller	1/0 or smaller	8	6
1 or 1/0	2/0 or 3/0	6	4
2/0 or 3/0	4/0 or 250 MCM	4	2
Over 3/0 to 350 MCM	Over 250 MCM to 500 MCM	2	1/0
Over 350 MCM to 600 MCM	Over 500 MCM to 900 MCM	1/0	3/0
Over 600 MCM to 1100 MCM	Over 900 MCM to 1750 MCM	2/0	4/0
Over 1100 MCM	Over 1750 MCM	3/0	250 MCM

SOURCE: Reprinted with permission from NFPA 70-1993, National Electrical Code.

path is adequate. The current-carrying capacity also needs to be derated when the grounding conductor is installed with multiple conductors in a conduit or raceway. Finally, when a single grounding conductor is used for several circuits, it should be sized for the largest overcurrent device protecting the conductors in that cable.

Of course, a conductor sized by the book does not always mean the grounding is properly accomplished, especially if it is a direct ground. The cable must be securely attached if it is grounding to structural steel (preferred) or to metal pipes in lengthy contact with the ground. It should be welded (preferred) or bolted using a connection with an area of contact greater than the size of the wire. There are several ways to accomplish the actual connection to the earth.

A common way is to use water or gas piping as the ground. (Metal electrical conduit should never be used even if it is buried or in the slab.) There should be a long enough section of pipe in contact with the ground, a minimum of 10 ft, for the rated voltage and amps. This can be a potential risk because it exposes maintenance personnel to electrical hazard.

It is odd to think of the earth as anything other than a "ground." This is not always the case. Moist soil can be a fairly good conductor. Electric fences for cattle actually use the ground as the return leg of the charging circuit. Even though there is a large voltage drop (especially when the ground is dry), enough current flows to make it an effective circuit. The same can occur in any grounding application, unless it is designed carefully and cautiously.

Structural grounding has a risk to maintenance personnel, similar to pipe grounding. The ground connection(s) should be limited to elements that are least accessible to maintenance activities. The structure must also be sufficiently grounded to carry the rated voltages and ampacities for the connected equipment.

Other grounding methods are grounding rods, a ground ring surrounding the building, or concrete-encased electrodes. Grounding rods should be at least $5/8$ in in diameter, iron or steel, and driven at least 8 ft vertically into the earth or buried in a trench at least 2.5 ft deep. Ground loops should be the same depth and at least 20 ft of bare copper conductor no smaller than #2 wire. Concrete-encased electrodes must be surrounded by at least 2 in of concrete and consist of at least 20 ft of #4 or larger conductive metal such as galvanized steel reinforcing rods. These are all minimum requirements and should be exceeded whenever possible.

In locations with dry soil, each of these methods can benefit from techniques to maintain the moisture of the ground connection. Some designers prefer use of a hygroscopic material that absorbs moisture from the atmosphere to keep the ground damp; others do it the old-fashioned way by locating an equipment condensate drain nearby. If the water volume is great, a "French drain," or small gravel filled pit, helps to percolate the water into the ground effectively. Such a wet ground ensures a positive connection for complete grounding. It also minimizes any potential hazard from ground conductance.

This discussion about grounding applies to all systems, including panels and all major equipment items. In the case of panels, it is a good practice to have a grounding bus in the panel, with an individual grounding wire for each panel according to the above criteria.

Overload Protection

Most reputable manufacturers have more than adequate protections built into their equipment. Even so, there is no harm in writing into the specifications a clear definition of the code, and project, requirements. In addition, many applications are custom designed for given installations. Unless specific overload protections are specified, they will not be provided. Then, when the equipment burns out because of a short or bad control element, there is a liability issue—or, in some cases, a life safety or health code issue. It is best to be conservative and to specify exactly what is needed, to keep everyone out of trouble. Table 9.3 contains the code requirements for just about any installation possible, including pumps, motors, and hvac equipment.

TABLE 9.3 Running Overload Units

Kind of motor	Supply system	Number and location of overload units
1-phase ac or dc	2-wire, 1-phase ac or dc, ungrounded	1 in either conductor
1-phase ac or dc	2-wire, 1-phase ac or dc, one conductor grounded	1 in ungrounded conductor
1-phase ac or dc	3-wire, 1-phase ac or dc, grounded neutral	1 in either ungrounded conductor
2-phase ac	3-wire, 2-phase ac ungrounded	2, one in each phase
2-phase ac	3-wire, 2-phase ac, one conductor grounded	2 in ungrounded conductors
2-phase ac	4-wire, 2-phase ac, grounded or ungrounded	2, on per phase in ungrounded conductors
2-phase ac	5-wire, 2-phase ac, grounded neutral or ungrounded	2, one per phase in ungrounded phase wire
3-phase ac	Any 3-phase	3, one in each phase

SOURCE: Reprinted with permission from NFPA 70-1993, National Electrical Code.

Chapter 10

Motors

There are not many motors on a medium-sized project, but they are usually quite important. They pump water, move air, and perform all the other vital functions needed to maintain a safe and comfortable environment in the building. The electrical engineer does not traditionally have much input to the selections of these motors because they are part of pumps, air handlers, or cooling towers. However, a diligent electrical designer should take an active role in the selection of motor-driven equipment for several reasons:

- To match the voltage and phase to the available service and to lobby for three-phase service at the highest practical voltage
- To investigate the quality, efficiency, and service history of the brand, style, and model of motors specified

This opportunity to review the motor selections is contingent on the cooperation of the hvac and plumbing designers. Usually they simply pass on the cut sheets of the selected equipment and expect the electrician to power them up with no further ado. Usually the selections are passed along near the end of the construction documents phase, and the electrician can do no more than just that. With just a little planning, though, proper equipment can be selected and installed. This section describes a few of the important design considerations.

Generator Sizing

There are many subtle characteristics of motors and their effect upon the power grid of a building. One of the least-appreciated aspects is the starting current, or kilovoltamperes, for large motors. A good way to get a feel for this portion of motor loads is to evaluate all of the things that go into sizing an emergency generator for a facility.

Any sizable building has an emergency generator on site, if only to handle life safety and fire apparatus at the facility, to power the following:

- Exit, stairway, and corridor lighting
- Fire alarms, panels, and other devices
- Pumps serving the building sprinkler or standpipe system
- Computer or communications systems
- Sump and sanitary waste pumps
- Elevators
- Automatic doors

Some facilities have a much higher load than others, such as hospitals or high-rise buildings. Others have a minimal requirement that can be handled adequately by battery packs in lights and alarm panels.

There may be several large motors on emergency power, such as elevator motors, fire pumps, stair pressurization, and associated smoke control fans. The problem with these large motors is that they can draw up to 10 times their full load running current when starting. This is illustrated in Table 10.1. If there are several such motors on a project, the generator will have to be greatly oversized.

If the code for a motor is not known, a reasonable assumption is to use the following criteria:

TABLE 10.1 Motor Starting Kilovoltamperes and Full-Load Running Kilowatts per Horsepower

Hp	Starting kVA						Running kW
	Code E	Code F	Code G	Code H	Code J	Code K	
1	4.7	5.3	5.9	6.7	7.5	8.5	1
2	9.5	10.6	11.9	13.4	15.1	17	1.9
3	14.2	15.9	17.9	20.1	22.6	25.5	2.8
5	23.7	26.5	29.8	33.5	37.8	42.5	4.6
7.5	35.6	39.8	44.6	50.3	56.6	63.7	6.9
10	47.5	53	59.5	67	75.5	85	8.8
15	71.2	79.5	89.2	100			13
20	95	106	119	134			17.2
25	118.7	132	149	168			21.6
30	142	159	179	201			25.5
40	190	212	238	268			35.2
50	237	265	298	336			43.5
60	285	318	357				51.5
75	356	398	446				63
100	475	530	595				84

Horsepower	Code
2 and below	M
3–5	K
7.5–10	H
15 and up	G

Consider the following loads on the emergency power system of a small multistory office building:

	Start up (kW)	Running (kW)
Lighting and exit signs	10	10
Elevator motor (10 Hp)	60	9
Stair pressurization fans (2 at 15 Hp)	89	13
	89	13
Totals	248	45

This is a fairly basic emergency power system. If the motors were started at full load, a 250-kW generator would have to be installed, even though it would only have a 45-kW load once the motors are up to operating speed. Usually the large motors can be staged. For example, one pressurization fan would be started by the fire alarm system, then the second after the first is at full speed. The controls could also stage the elevator motor in this sequence, if the code regulating authority allows it. This would reduce the generator capacity to the value of the resistance loads plus the start-up load of the single largest motor.

Lighting and exit signs	10 kW
Elevator motor	9 kW
Stair pressurization #1	13 kW
Stair pressurization #2	89 kW
Total	121 kW

This reduces the generator size by half, but it is still loaded at less than 40 percent capacity. This is good for the generator itself because the windings will heat up less and last longer. However, the prime mover—the gasoline, diesel, or natural gas motor—does best at a 50 to 90 percent loading. Because this is the major maintenance item in the whole system, loading it so low can be a problem in the long term.

Reduced voltage starting

A compromise is to size the generator to permit a drop in voltage to the emergency power grid. Table 10.2 is an approximate sizing chart for

TABLE 10.2 Motor Starting and Maximum Starting Kilovoltamperes (Three-phase, 240-V Wye)

60-Hz standby	Frame size	Voltage dip (%)				
		15	20	25	30	35
12	279	13	18	23	28	34
17	280	16	21	28	35	43
22.5	281	20	31	38	54	59
30	282	23	41	45	65	72
35	283	33	47	60	76	90
40	284	40	57	73	92	110
45	285	48	69	88	110	132
50	286	50	70	91	115	134
55	287	64	91	116	145	172
60	288	64	95	124	159	190
66	289	78	109	134	163	190
80	330	100	140	179	215	255
100	351	112	159	200	250	295
125	371	150	207	270	323	380
150	390	170	242	310	388	460

this practice. The starting kilovoltamperes for the largest 15-Hp motor, at a reasonable voltage dip of 25 percent, is only about 25 kVa. This is much better than the 89 kW required for full voltage starting, and it means that the generator has to be oversized by only about 12 kW to handle the full load.

Many motors will be started unloaded, such as fans, pumps, and motor-generators. A 30 percent voltage dip is acceptable for such devices. However, if the drop is much more than 30 percent, the motor starting contactors and control relays may not stay closed long enough to start the motor. Starting loaded motor circuits should be limited to 20 percent voltage dip. Very sensitive circuits may need to be limited to even less of a voltage variation.

A popular alternative is the use of soft starters for motors. These accomplish reduced voltage starting electronically, and they quickly bring the motor up to speed. They should be specified for any large motor that has several starts and stops during the course of a day. The soft starter is just as economical as a much more costly variable-speed drive in situations where the load is full or close to full load, when the motor is operating. Often this can be confirmed during the commissioning process or sometimes not until the fan or pump motor is installed and the building is functioning on a regular occupancy schedule. Often motors installed with a VFD operate on a fairly uniform load the majority of the time. If this can be verified during commissioning, or in existing buildings as part of an energy audit, installing a soft-start motor in lieu of the VFD can result in considerable savings in installation cost and long-term maintenance cost as well.

Derating

Other unique considerations apply to single-phase motors started on a three-phase generator. There are three possible situations:

1. If several single-phase motors are started so that they present a balanced load on all three phases, the generator sizing procedure is unchanged.
2. If the entire load is single phase and the generator is single-phase connected, only two-thirds of the generator windings are used. Thus, only two-thirds of the rated figures in the tables apply.
3. If a large single-phase motor load is applied to one phase of a three-phase generator (phase-to-ground) or two phases, only one-third of the windings are effective on a Wye-connected generator. This can be as little as one-sixth effectiveness on a Delta-connected generator.

This is why larger motors, especially if they are to be on emergency power, should be three-phase connected.

Another type of load to be wary of are those comprised of silicon controlled rectifiers (SCR), or thyristers. These cause generator stator heating and interfere with the voltage regulation system. Both effects require generator derating and voltage regulator filtering. Some common SCR loads are uninterruptable power supplies (UPSs) used to back up power to computer systems, reduced voltage motor starters, and VFD motor speed controls. Even though each of these items reduces the actual load on the emergency system, the adverse effect on the quality of the power supply partly offsets this benefit. Not only must the generator be oversized to offset these effects, but expensive power filtering apparatus must also be added.

Two more derating factors are altitude and temperature. At high altitudes the air is less dense, so there is less cooling of the generator and less heat extracted from the engine combustion processes. As ambient temperature increases, the cooling is also reduced because the temperature difference is less. Conservative values are a reduction in rating of 1 percent for each 250-ft increase in elevation or 10° increase in temperature. There is also a derating for fuel that in some cases must be taken into consideration.

Peak building loads

This discussion of generator sizing illustrates the high starting current needed to get a motor up to speed. Table 10.3 gives the locked rotor current, which is the current at start-up under full load. This is quite high, compared to the full load amps given in Tables 10.4, 10.5, and 10.6. This effect is not only important in the sizing of the building

TABLE 10.3 Maximum Locked Rotor Current (Three-Phase Motors)

hp	200 V	220/230 V	440/460 V
1/2	23	20	10
3/4	29	25	12.5
1	34.5	30	15
1 1/2	46	40	20
2	57.5	50	25
3	73.5	64	32
5	106	92	46
7 1/2	146	127	63
10	186	162	81
15	267	232	116
20	334	290	145
25	420	365	182
30	500	435	217
40	667	580	290
50	834	725	362

SOURCE: Reprinted with permission from NFPA 70-1993, National Electrical Code.

emergency generator but in evaluating the overall energy use characteristics of the facility. The most marked impact is on the demand load billing of the facility.

Every time the power to the building drops off-line and then comes back on all the motors start up simultaneously. This can cause a very high peak demand, quite out of proportion to the usual operating load. A minimal energy management system, to stage the major motors back on-line after a brown-out and first thing in the morning when the facility is occupied, can greatly minimize this peak. In fact, using an emergency generator to handle this peak until the main service is back on-line can pay for itself very quickly in areas that have frequent power outages.

This introduces another possible use for a building emergency power system. This is very expensive equipment to purchase and keep maintained, only to use it a few times a year, if that. Both the prime mover and the generator are designed and constructed for continuous operation for many years. The emergency response of the system is not diminished if the system is operated on a regular basis, as long as the proper maintenance is done per the manufacturer's recommendations. In fact, quite the opposite may be true. Using the emergency generator for peak load reduction can save a great deal on the utility bills—enough so that the system can be installed on smaller, otherwise marginal facilities.

Even if this peak load reduction system is not to be installed at a facility, a little extra planning makes it very easy to configure it into the power distribution system in the future. As energy costs escalate and building operators strive to reduce consumption, this will be an easy fix

TABLE 10.4 Three-Phase ac Motor Data

Motor Hp	Motor volts	Motor amps	Size breaker	Size starter	Heater amps	Size wire*	Size conduit
1/2	230	2	15	00	3	12	3/4
	460	1	15	00	2	12	3/4
3/4	230	2.8	15	00	4	12	3/4
	460	1.4	15	00	3	12	3/4
1	230	3.6	15	00	5	12	3/4
	460	1.8	15	00	3	12	3/4
1 1/2	230	5.2	15	00	8	12	3/4
	460	2.6	15	00	4	12	3/4
2	230	6.8	15	0	9	12	3/4
	460	3.4	15	00	5	12	3/4
3	230	9.6	15	0	12	12	3/4
	460	4.8	15	0	7	12	3/4
5	230	15.2	15	1	20	12	3/4
	460	7.6	15	0	10	12	3/4
7 1/2	230	22	40	1	30	10	3/4
	460	11	30	1	15	12	3/4
10	230	28	50	2	35	10	3/4
	460	14	30	1	20	12	3/4
15	230	42	70	2	50	8	1
	460	21	40	2	25	10	3/4
20	230	54	100	3	70	6	1
	460	27	50	2	35	10	3/4
25	230	68	100	3	80	4	1 1/2
	460	34	50	2	40	8	1
30	230	80	125	3	100	3	1 1/2
	460	40	70	3	50	6	1
40	230	104	175	4	150	1	1 1/2
	460	52	100	3	70	6	1
50	230	130	200	4	175	0	2
	460	65	150	3	80	4	1 1/2
60	230	154	250	5	200	000	2
	460	77	200	4	100	3	1 1/2
75	230	192	300	5	250	0000	2 1/2
	460	96	200	4	125	2	1 1/2
100	230	248	400	5	300	350 mcm	3
	460	124	200	4	175	0	2
125	230	312	500	6	400	500 mcm	3
	460	156	250	5	200	000	2
150	230	360	600	6	450	700 mcm	4
	460	180	300	5	225	0000	2 1/2

*Wire size may vary depending on type of insulation and other factors.
SOURCE: Reprinted with permission from NFPA 70-1993, National Electrical Code.

if the original system is well designed. The keys to this forward planning are (1) large motors contribute most to the peak load, (2) single-phase motors make it harder to take advantage of the full generator capacity, and (3) solid state SCR loads should be minimized because of their adverse effect on the quality of the power supply. An economical solution is to design the building power grid with a separate equipment

TABLE 10.5 Full-Load Current in Single-Phase ac Motors

Hp	115 V	200 V	208 V	230 V
1/6	4.4	2.5	2.4	2.2
1/4	5.8	3.3	3.2	2.9
1/3	7.2	4.1	4	3.6
1/2	9.8	5.6	5.4	4.9
3/4	13.8	7.9	7.6	6.9
1	16	9.2	8.8	8
1 1/2	20	11.5	11	10
2	24	13.8	13.2	12
3	34	19.6	18.7	17
5	56	32.2	30.8	28
7 1/2	80	46	44	40
10	100	57.5	55	50

SOURCE: Reprinted with permission from NFPA 70-1993, National Electrical Code.

TABLE 10.6 Full-Load Current for Two-Phase ac Motors (Four-Wire)*

	Induction type squirrel cage & wound rotor					Synchronous type unity p.f.			
HP	115 V	230 V	460 V	575 V	2300 V	220 V	440 V	550 V	2300 V
0.5	4	2	1	0.8					
0.75	4.8	2.4	1.2	1.0					
1	6.4	3.2	1.6	1.3					
1.5	9	4.5	2.3	1.8					
2	11.8	5.9	3	2.4					
3		8.3	4.2	3.3					
5		13.2	6.6	5.3					
7.5		19	9	8					
10		24	12	10					
15		36	18	14					
20		47	23	19					
25		59	29	24		47	24	19	
30		69	35	28		56	29	23	
40		90	45	36		75	37	31	
50		113	56	45		94	47	38	
60		133	67	53	14	111	56	44	11
75		166	83	66	18	140	70	57	13
100		218	109	87	23	182	93	74	17
125		270	135	108	28	228	114	93	22
150		312	156	125	32		137	110	26
200		416	208	167	43		182	145	35

*For 90 and 80 percent power factor multiply the above figures by 1.1 and 1.25, respectively.
SOURCE: Reprinted with permission from NFPA 70-1993, National Electrical Code.

riser, as is done in hospitals, on which only large motor loads are connected. If it is absolutely necessary to power single-phase motors, they should be balanced on this riser. Otherwise, all motors should be single-speed, three-phase, high-efficiency models.

These are some very important considerations for the life cycle costs of operating a facility. It is rare for the electrical designer to have much input, though, because the major equipment items are selected by other disciplines, according to their own unique criteria. This is not a good practice because the building should be looked at as a single integrated system when it comes to electricity usage. For example, the hvac engineer can accomplish the same fan and cooling savings with either a single- or three-phase motor in an air handler. The power source makes no difference to cooling capacity.

It is a completely different picture for the electrical side of the system when peak loading, generator sizing, and peak load reduction are taken into consideration. Using the emergency generator to handle a part of the peak cooling load each afternoon in the summer can pay for the whole emergency power system in a couple of years or less. The payback is even swifter if the cost of battery packs for lights is taken out of the budget (because these lights will be circuited separately, on the emergency circuits). As a result the building will not only be more energy efficient but also safer and more dependable for the occupants. This translates into lower insurance costs for the owner and higher rental rates, especially if the emergency system can help to keep sensitive computers on-line during an outage. Each business will need a small UPS for their server at the instant power is lost. After that, being able to continue to use the office computers would be a great benefit to almost any modern business.

Motor Coordination

The most important point in the previous discussion is that when a motor is started, there is a high inrush current, up to 10 times the rated full-load amps. The true significance of this fact is lost to most mechanical designers unless they have been involved in energy audits and utility bill analysis.

Take a common example: An air-cooled chiller with eight fans that can be cycled on and off according to the load. The control schemes are usually designed to minimize the energy use or the run time of the fan. If two or three fans can be turned off even for an hour or two, that is what happens. No thought is given to what happens when the three fans must turn on at the end of that time. The cooling load has peaked by this time, and the condensers and air handler fans in the building are increasing in speed, simultaneously. Concurrent with this, the chiller fans cycle on. All told, it is possible for five or six fans to start at the same time, causing a substantial spike in the building electric demand rate.

Depending on the rate schedule, this spike will affect at least the billing for that month and usually the month or two on either side—sometimes the whole year's billing. It often only takes one such peak every 12 months to ratchet the whole utility bill upward. Controls can be configured to start the motors in a random sequence. They can also be programmed to keep the motor running when it might otherwise be cycled off, if the effect on the peak load is less than that on the regular usage charges. The extra fan energy is never lost because the fluid or air will still benefit from the cooling effect, from either a fan or compressor. This is especially true for large buildings, which can benefit from thermal storage effects of mass (which store any overcooling energy for future use). So, the penalty is strictly due to demand billing charges.

Another common circumstance is a facility with several rooftop units. Often they are on a time clock, which cycles the equipment on an hour or so before opening for business. This, of course, causes a giant peak every day. The same occurs whenever there is a power outage, and all units start up simultaneously when power is restored. A simple random start relay at each unit will alleviate this problem.

The key to planning each of these situations is an understanding of how peak load is billed. First, the demand charges are typically only levied during regular business hours. Second, the peak demand is measured as the average power usage during a 15-minute interval. Third, demand charges are typically only revised once a year, so a high peak will affect billing for an entire year. Some would consider the motor starting current, which can be 6 times the running current, to be invisible to peak demand meters because it lasts only a few minutes. Still, if it draws 100 A more for 3 min, this averages out to 20 A more for the 15-min peak demand period. The situation is exacerbated considerably when many motors are started at once, as after a power outage. Even though the power may be lost only once in a year, this is enough to increase peak demand charges for the next 12 months or more under most commercial rate schedules. I know of no facilities that have gone a year or more without an outage; even one can justify the cost of an emergency power generator supplying the equipment branch or, at least, building controls for this branch.

Motors in a Mechanical System

Mechanical designers are increasingly specifying variable-speed drives for all the major motors at a facility. Instead of starting and stopping motors, the speed is modulated uniformly over the range of speeds.

This is not done strictly to minimize peak loads (which it does quite well) but to take advantage of the fan laws. All other factors being constant, the horsepower input to a fan varies as the cube of the speed:

$$\text{Hp}_2 = \text{Hp}_1 \times \frac{N_2^3}{N_1^3} \quad (10.1)$$

where N = speed of the motor, r/min. This means that if the speed is increased by 10 percent, horsepower increases by a factor equal to

$$\frac{\text{Hp}_2}{\text{Hp}_1} = \frac{1.1^3}{1.0} \quad (10.2)$$

This is a 33 percent decrease. Reducing the speed by 20 percent results in an almost 50 percent savings in input horsepower.

This is the driving force behind the move to install VFDs on all major motors that can reasonably be modulated at a facility. The only problem is that the solid-state drives cause a lot of harmonics in the whole power system, which reduces the overall efficiency of the entire power grid. Of course, this is of no concern to the mechanical designer, mostly because it is such a difficult effect to quantify.

What the hvac designer can understand is that a two-speed motor costs a fraction of a VFD, and saves almost as much energy as a VFD. The controls are simpler and more dependable to configure, operate, and troubleshoot. Also, a clever control scheme can use the extra cooling or heating energy that is generated when the motor is rotating the fan faster than is really necessary, to balance when the opposite is the case. Finally, many mechanical systems naturally operate at very near full speed most of the time anyway and can be designed to operate at near half speed most of the heating season (or whichever is the lesser of the seasonal loads). In such situations the VFD does very little actual good, other than to make it extremely easy for the hvac designer to design and specify the system.

The electrical designer, on the other hand, has to cope with the harmonics. Worse still is the great extra cost added to the electrical budget on the job. The VFD itself costs quite a lot more than a two-speed motor, and if the motor is on emergency power, the generator must be oversized as well.

These are just a few of the reasons why the electrical designer should question each use of VFDs on a job. Some applications can perform almost as well with two-speed motors instead. The important thing is to look at the life cycle cost of the installation.

The initial cost of a VFD is actually far more than the drive itself. Varying the air flow at an air handler, for example, requires a far more complicated hvac system. Variable air volume boxes (sometimes fan-powered ones) with reheat coils (sometimes electric) are needed for each zone, with expensive pneumatic or electronic controls for each device. Each of these devices adds cost and complexity to the job. With complexity there is greater potential for the system to operate out of adjustment and to use more energy.

Variable speed pumping has the same ramifications. Additional valves, controls, and balancing devices are needed, and sometimes even a completely new piping loop. As before, it is a very complex system that has many ways to go out of adjustment. Like the engine in a fancy sports car, the system is extremely efficient when it is tuned up properly; otherwise it can be a maintenance nightmare and waste a lot of energy.

These sophisticated systems can add a great deal of cost to the entire project budget, not only to the mechanical but also to the electrical budget because of the cost of the VFD and associated controls. It would be a disservice to the client to specify such an installation if it cannot be justified on the basis of savings on the utility bills. This analysis should take into account the maintenance aspect as well, both the extra annual cost in having to have a more highly qualified staff and the cost of the repairs themselves.

Often a simple but rugged system suffices. A few very basic but dependable control devices, such as two-speed motors, a time clock, random start relays, and override timers, can often save almost as much energy as a fully variable system, at a fraction of the cost.

Operating Parameters

Mechanical systems are complex. They are different from electrical systems. Everything is variable: flow, pressure, static head, dynamic head, friction losses, temperature, density, altitude, and humidity, to name but a few of the possible variables. The selection of the driving force for the motor is therefore quite difficult. The motor and its drive must meet varying conditions throughout the operating range of the system. These conditions are characterized by two important parameters, horsepower and torque.

Horsepower

The fundamental definition of horsepower is the time rate of energy use that results in motion (i.e., work). One horsepower of work is the amount of power needed to raise 33,000 lb · ft in 1 min. This is expressed mathematically as

$$\text{Hp} = \frac{\text{load} \times \text{feet per minute}}{33{,}000} \qquad (10.3)$$

As an example, if an elevator weighs 6600 lb fully loaded and the motor is designed to raise the carriage 100 ft in 30 sec, a 40-Hp motor is required.

When an elevator motor is started, it has a dead load to lift, and quickly. Even if the lift is not loaded to capacity, the motor must still develop sufficient torque to lift it and to overcome the friction of pulleys, rail guides, and other parts of the whole mechanism. This requires the most torque or instantaneous power of any motor application. Even a VFD must power up fast, so its ability to lower peak load and diminish inrush current is limited. A mechanical flywheel would do far more good because it can store the potential and kinetic energy of the elevator in a rotating mass. When the elevator starts again, this mass releases its energy instantaneously to the lift, sparing the peak inrush load on the motor. This idea is developed mathematically later in the chapter, in a discussion of inertia, or rotational momentum.

The principles are not quite as neat and easy to understand for fans and blowers, but the same general idea applies. The value is still a load, defined as a static pressure the fan is working against times a rate divided by a constant or proportionality:

$$\text{Hp} = \frac{\text{Psf} \times \text{Cfm}}{33{,}000 \times \text{Eff}} \tag{10.4}$$

where Cfm = cubic feet per minute of air flow
Psf = static pressure in lb/ft^3
Eff = efficiency

The concept of efficiency is introduced here. The significance is that as efficiency decreases, the horsepower needed to drive the system increases by the same proportion. Any mechanical system has inefficiencies because motor power is lost to heat and friction, and there is a nonuniform power supply.

There are some other interesting consequences of Eq. 10.4 when you consider the possibility of starting the fan at reduced horsepower. In a duct system there is much less static friction if the airflow is reduced, so a variable volume system benefits from a reduction of both parameters. At start-up, the fan in a ducted system does not initially see the full static pressure because air is a compressible fluid. Once the duct system, the building, and the return air plenum or ducts are pressurized, the static pressure reaches the design level. The full horsepower of the fan is then needed to move air all the way through the distribution system. This process can take a few minutes in a large system, so the peak of inrush current for the motor is fairly long lasting.

Other fans are used in applications where volume is more important than flow against a static pressure or pressure head. Cooling tower fans and attic or room ventilators are typically large propeller-type devices that move a large volume of air against a relatively small static pressure. These devices can usually be started slowly for a nominal cooling or exhausting effect. This is often beneficial in the application, as it is with cooling towers. These devices are well suited to VFD installations, and they usually result in a lower inrush as well as peak current usage.

The horsepower for the third important motor application—pumps—is, again, a function of a rate, a pressure, and a constant:

$$\text{Hp} = \frac{\text{ft} \times \text{gpm} \times \text{sg}}{3960 \times \text{Eff}} \qquad (10.5)$$

where ft = feet of head
gpm = flow in gallons per minute
sg = specific gravity (water = 1.0)

This application is further complicated by some unique characteristics of pumps for centrifugal designs: The efficiency varies with the flow rate. Some estimated efficiencies are as follows:

500 to 1000 gpm	70–75 percent
1000 to 1500 gpm	75–80 percent
Over 1500 gpm	80–85 percent

This means that there is a penalty as flow is reduced. Although the flow is diminished, and the necessary driving horsepower with it, the efficiency drops, which requires more horsepower. The efficiency of displacement pumps varies between 50 and 80 percent.

The performance of pumps is different from either the constant load elevator motor or the fan motor. It has characteristics of each because the fluid is more viscous and because the fluid has a fixed static head. This head is due to the change in elevation or having to pump water many feet above the plane of the pump. As the head increases, the starting horsepower increases, and the system behaves more like the fixed force needed to start the elevator. Conversely, in a single-story building the static head may be low, and the pump motor will need less starting power to get the water moving. In the first case a variable drive pump may start at a high horsepower anyway, and the benefits common for such devices would be diminished.

Torque

These applications have introduced the implications of starting load on a motor, due to the specific application. This phenomenon is best explained by introducing an important concept: torque. Torque is, fundamentally, a function of the force applied and the length of the arm by which it is applied. In mathematical terms,

$$\text{Torque} = \text{force} \times \text{distance} \tag{10.6}$$

Torque is measured in ft · lbs. If a 100-lb force is applied at the end of a foot-long wrench with a nut in the grips, the torque applied to the nut is 100 ft · lbs. For radial systems, the function becomes

$$\text{Torque} = \text{force} \times \text{radius} \tag{10.7}$$

The torque is still measured in foot pounds, but the distance is not the radius of the shaft or cam. If a cam 1 ft in radius is able to lift a 100-lb weight, the torque applied is 100 ft · lbs.

This introduces the idea of torque in a rotating dynamic system, which is defined as

$$\text{Torque} = \frac{\text{Hp} \times 5250}{N} \tag{10.8}$$

where N = speed of the motor, r/min. Rewriting the equation in terms of horsepower,

$$\text{Hp} = \frac{\text{torque} \times N}{5250} \tag{10.9}$$

Now you can see the important benefit of starting a motor at reduced speed. For the same amount of torque applied to the load, less horsepower is required. Sometimes, though, a reduced speed does not do much good. For example, a pump pushing fluid against a high static head (due to a column of water as well as flow resistance due to valves and rough, pitted pipe) may not accomplish anything at low revolutions per minute. A fan may move air but perhaps not through the full extent of the ductwork. A motor acting against a heavy fixed load such as an elevator may be below the threshold to get the elevator started moving (i.e., to overcome weight and friction) at low revolutions per minute. If so, the motor will just burn up and the elevator will get nowhere.

Another common instance of this kind of phenomena is when there are two pumps in parallel. Sometimes only one of the pumps is put on a VFD, with the intention of cycling the pumps according to the full capacity when both pumps are operated. What can happen is that the first pump on the VFD must start at a high speed just to overcome static and dynamic head and friction forces. Then, as load increases the VFD pump is cycled off and the full-load pump is brought on-line by itself. As the load increases still further, it is necessary to bring the first VFD pump back on-line. Now it must operate against an even higher pressure and starts at nearly full load. So, it's almost better to have no VFD at all, unless separate drives are provided for each pump.

Accelerating torque for rotary motion

As the required response time decreases, a higher torque drive must be installed. The acceleration available for a given torque is determined by the function

$$t = \frac{W \times K^2 \times \text{delta-}N}{308 \times T} \qquad (10.10)$$

where t = time
WK^2 = inertia
delta-N = change in speed, r/min
T = torque

Inertia, in this context, is a measure of the dynamic mass of the system. Of the systems considered, the air handler fan sees the least amount of inertia, then the fluid pumping system, followed by the elevator motor lifting a dead weight. For any given application, a higher torque is needed to accomplish a greater speed change in a fixed amount of time. As the inertia of the system increases, the torque required to accomplish this task in a given response time increases.

The modern, sophisticated systems in buildings often have ways to reduce the starting load on a system, without resorting to the artificial load reduction of the motor on a VFD. For example, when a large hydronic pump starter is brought on-line, valves can be closed to minimize the fluid loop so that the starting pressure head is low. As the fluid is brought up to design flow, branches can be opened, air handlers brought on-line, and so forth. This is typically the starting sequence anyway because the load on the central chiller must be staged so that its peak is not reached right away. Warm-up cycles operate in this way as well.

Air systems are much the same way. They can be configured and sequenced to minimize the starting load on the fan by closing off branches and VAV boxes and short circuiting most of the air right back to the return air pathway. As the load on the fan diminishes, the branch ducts can be supplied sequentially until the entire system is up to operating speed and capacity.

All of this can be done without the benefit of a VFD on the motor (the elevator equivalent to the analogy was previously described as having a flywheel). Thus, the initial inrush peak is avoided by using existing system controls. In the case of air conditioning, the equivalent of a flywheel can be used to phase out VFDs all the way. The entire system can be either all on or all off, using the thermal storage capacity of the building to cool or heat the building when the system is all off.

These discussions indicate that a VFD-driven motor may be a redundant system when it is installed in a facility with a full energy management system (EMS). The EMS has capabilities that can minimize peak inrush currents, one of the two reasons for VFDs. The EMS can accomplish further savings by reducing static pressure or head, which diminish motor horsepower. Thus, all the VFD does is reduce motor revolutions per minute. Although this will certainly save on electricity usage, the payback on the VFD installation is greatly increased, perhaps beyond the minimum 7-year payback.

The bottom line is that VFDs should not be specified for all large motors indiscriminately. Each application must be studied carefully and a life cycle cost analysis done to justify the high initial cost of the variable frequency drive installation and all of the additional electronic controls needed to use the VFD to its full potential.

Inertia

Inertia is a physical quantity that is closely related to the more commonplace term *momentum*. It is the property of a body such that it will continue in motion after the motivating force is stopped. A car continues to roll if the engine is put in neutral while moving at high speed. A motor continues to rotate after the power energizing the coils is turned off.

The simplest inertial object is a rotating shaft. Its inertia is defined by the factor WK^2. This is the multiple of its weight in pounds and the square of its radius of gyration in feet. For a solid cylinder, the latter is defined by

$$WK^2 = 0.000681 \times \rho \times L \times D^4 \qquad (10.11)$$

where W = weight in pounds
ρ = density of the material
L = length of the shaft
D = diameter

Notice that the diameter is the most influential factor in this analysis. The further the mass is located from the axis of rotation, the higher the inertia of the object. This is why large flywheels—which are designed to store energy in the form of rotational motion by virtue of inertia—have a large mass away from the axis of rotation. There is only sufficient mass and structure inside for structural and dynamic support of the outside mass.

Objects with a high inertia not only store a lot of energy, they also require a lot of energy to get started. Flywheels, for example, are hard to get started rotating. Conversely, they can discharge their energy almost instantaneously. This can be used to create very high torques at the shaft, which in some cases is transformed into electricity of very high voltages.

Systems possess inertia as well. In a closed fluid loop of an incompressible fluid such as water, for example, all of the fluid must be brought up to speed simultaneously. The pump motor must have a high starting torque to overcome the frictional forces, static head, and inertia of the fluid before flow can be initiated through the loop. This peak current draw coincides with the peak starting inrush current of the motor, resulting in a prolonged, high-peak demand.

This same process happens whenever the pump is stopped. Each time it restarts, the inertia must be initiated again, the frictional forces overcome, and fluid head counterbalanced. It is better to have a VFD on the pump motor to keep the fluid circulating at a nominal rate sufficient to balance these forces. Then, as demand increases and a higher water flow rate is called for, the flow can be increased easily.

Air systems are more flexible because air is a compressible fluid. The entire volume must not be in motion before substantial flow can be initiated. Consequently, the starting torque is less than for a hydronic system. Nor is there as much of a penalty for bringing the fan to a full stop. Low airflow rates see a lower static pressure, and the system requires less static pressure at the fan to operate. In some cases, air just does not reach the most distant diffusers in the system, but it still moves. This is acceptable as the system is starting up because as the fan gets up to speed enough static pressure is developed to push air to the most distant diffusers of the system.

These principles depend upon the installation. The piping or ductwork may be arranged so that the controls system can artificially reduce the load on the motor at start-up. In such cases, an expensive (in terms of both initial cost and operating cost) high-torque VFD is not necessary.

Two-speed motors

Operating a motor at reduced speed uses much less energy than at full speed, other factors being constant. Most mechanical systems have three fairly unique flow regimes: minimum, typical, and maximum. The operating times at each of these conditions are proportional to a bell curve distribution, with the minimum and maximum each required about 10 percent of the time. This leaves an operating standard that applies roughly 80 percent of the time. During this time the load on the system varies by ± 10 to 20 percent.

In most fluid systems the minimum flow may be small, but the horsepower used by the motor is substantially higher. It must overcome the complete static head and frictional forces for the whole system, in addition to moving the minimum amount of fluid, air, or water, as the case may be. Therefore the minimum is not too far removed from the typical power level.

Compare, given these common conditions, a VFD motor with a two-speed motor designed with the lower speed at the typical operating speed. Assume the motor operates at this constant speed at all times, with the exception of when it must operate at maximum horsepower, at which time the motor operates at full speed. Given a bell distribution, the load on the motor that falls below the typical and above it are equal. Operating the motor at the same speed in all these conditions uses the same amount of energy as a motor modulated to exactly match the load. If the space conditions do not fluctuate excessively, there is no advantage to using a VFD on the motor.

Sophisticated control systems are now configured to do many things to reduce peak load, cycle equipment off, and otherwise reduce demand and use of hvac systems. Such a control oversight makes the two-speed motor even more competitive because the building load is varied already by other means.

VFDs are also less efficient at low speeds. Modulating a motor down to the minimum required speed can potentially use just as much power as the two-speed motor on its low speed (which is actually higher than the minimum VFD speed). Likewise, there is always some inefficiency associated with driving a motor with a VFD rather than directly. The VFD motor at full speed will therefore use measurably more power than the two-speed motor at full speed. So, even if the latter must run a little longer to satisfy the building load, it will not only use the same amount of electricity but have a lower peak-demand use.

In the early years of energy management, VFDs were an attractive way to save energy. With the advent of inexpensive, efficient building controls that are used on all major systems in new buildings, the ability of VFDs to save as much energy is diminished. In addition, the tolerance of any given electrical system for harmonic distortions is limited. With the advent of widespread use of computers in virtually every work environment, the room for additional digital devices is greatly limited.

A final blow to variable frequency drives is the advent of high-efficiency motors. The more efficient the motor being controlled, the less savings accrue from the use of a VFD. A properly sized, premium efficiency motor uses substantially less energy than a standard motor, and with much less trouble in terms of controls and maintenance.

Keep in mind that the use of a single- or two-speed motor in an hvac application does not diminish the versatility of the air conditioning system. There are still many dependable ways to modulate air and water flow in ducts and pipes. Inlet vanes and bypass valves perform the task as effectively as varying motor speed, the only detriment being they use more energy in terms of fan horsepower. However, as noted above, this savings may not be sufficient to justify the VFD in terms of life cycle costing or even simple payback.

Pump Curves

This discussion on the applications of motors in hvac systems is not intended to make the electrical designer into a mechanical expert. A little understanding of the way large motors are used by mechanical systems will open the door for some electrical input into the specification of equipment. After all, the VFDs and controls are usually in the electrical budget; the alternative systems are not. If the electrician can show that the installation of simpler, more traditional systems will cost less to install and operate, that is what should be done. The only way to do so is if he or she fully understands the use and application of the motor-driven equipment.

The above is the strongest argument in favor of constant-speed applications, especially in smaller installations. It is difficult, in such circumstances, to justify the extremely complex and costly variable air volume or variable pumping systems. Initial cost is high, and maintenance is costly and requires expert personnel on staff. Small packaged systems have very competitive operating characteristics and many of the same controls.

Larger systems are another story altogether, such as for a hospital, industrial complex, or large school or college campus. The staff is usually competent in maintaining the complex controls of variable systems. Smaller packaged or built-up systems cannot compete with the efficiency of the central systems. The installation still requires a close inspection of the annual and daily hvac loads on the facility, as described previously. This is a valuable perspective, especially if there is data available from the owner on a similar, existing system—such as electrical billings, chiller electrical loads, or a thorough energy audit analysis.

There is another compelling argument that the mechanical engineer can use in favor of variable-speed drives. As with most such arguments, if it is presented from one perspective, it is almost impossible to overcome. However, there are strong factors against variable-speed applications, and it is likely that the only design team member who has the expertise to present contradicting evidence is the electrical engineer on the project. The evidence has to do with pump and fan curves, which describe the operating characteristics of the device in a particular design application.

Mechanical throttling versus VFD control

Most pumps used today are centrifugal pumps. Their operation is defined by two independent curves. The pump curve is a function of the pump impeller and motor only. The other curve is a system curve. This curve depends on the system in which the pump is installed, its flow requirements, static, and dynamic head. The two curves intersect at the natural operating point, where the pressure and flow provided by the pump matches the needs of the system.

Figure 10.1 illustrates the two alternatives. Notice how the available flow and head vary as the operating point changes in the two systems. As the throttled system is reduced in flow, the available head increases. As the adjustable speed control system is slowed down, the available pressure head decreases dramatically. There are several aspects of the hydronic system that have a relatively constant pressure requirement, regardless of flow rate:

- The static head of the system, by virtue of the change in elevation of system components, is constant.
- The static pressure needed to move fluid through coils is unchanged.

The components that require less pressure, due to a reduction in flow:

- Flow through the mains and branches (which see a minimal change if it is a reverse-return loop system, as is common in large installations)

The total pressure head required decreases very little. Even though many cooling coils may be isolated from the loop, the same amount of pressure must be developed by the pump to move the fluid, assuming

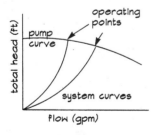
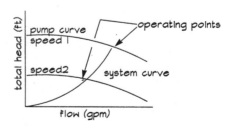

(a) curves for throttling valve flow control

(b) Curves for adjustable speed flow control

Figure 10.1 Typical pump operating curve.

all the coils are approximately the same size. Consequently, it is sometimes impractical to reduce flow using a VFD. Staged pumps, sized to provide the necessary head at different flow rates, offer a better alternative at a much lower cost.

This is a difficult concept to fathom. Compare it to a long lighting circuit, with all the fixtures in series. The same voltage will be needed to light one fixture whether it is the first in the loop or the last one. The same principle applies to hydronic systems, which are usually configured in series. There will be a little less friction lost in the pipe mains, but all other factors are unchanged. Figure 10.2 gives some more illustrations of these systems.

Centrifugal fans follow the same basic premise. A VAV system is designed to provide a constant pressure in the main supply duct, and the VFD varies fan speed according to a pressure sensor. The alternative is a constant speed fan with inlet vanes to vary the airflow through the unit. Notice in Fig. 10.3 that the system and fan curves are quite similar. Also notice that in Fig. 10.4 the power requirement for the VFD unit is substantially lower because the fan revolutions per minute are reduced.

There is one item that changes for VAV systems (Fig. 10.5). The system has a minimum static pressure. This pressure is needed for the VAV box controllers to operate properly. This requirement raises the system curve, indicating as before that a relatively constant pressure is needed at the fan discharge.

Mechanical designers, when trying to justify the use of VFDs, usually leave this last illustration out. It shows that the ideal system does indeed require less power, but the nature of the VAV boxes requires most of the work be put back into the system for it to operate properly. The result is a very complex and expensive air distribution system that cannot be justified on the basis of the energy saved.

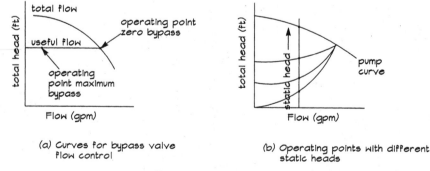

Figure 10.2 Pump curves at different static pressures.

Motors 165

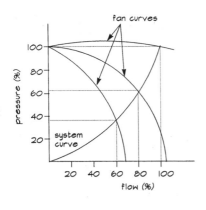
(a) Inlet vane settings

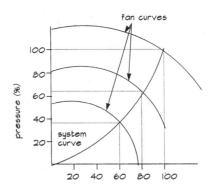
(b) VFD settings

Figure 10.3 Typical system curves and fan curves.

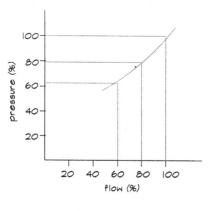
(a) Inlet vane settings

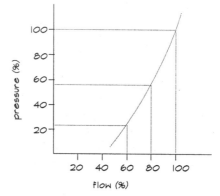
(b) VFD settings

Figure 10.4 Power settings versus flow.

Variable torque loads

As if all the analysis so far were not confusing enough, there is one last function to be aware of. The principles noted so far are based on the affinity laws. These establish, among other things, that motor horsepower is proportional to the speed cubed. A slight reduction in speed has a disproportionate effect on power requirements. This can justify a lot of controls and extra equipment from the energy saved.

Centrifugal fans, pumps, and blowers do not follow this cube relationship. They are all variable-torque loads, which require much lower torque at low speeds than at high speeds. The torque is proportional to the speed squared by the formula

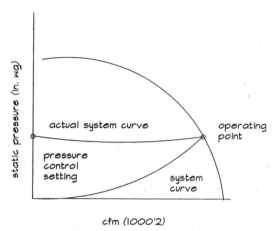

Figure 10.5 Variable air volume application with static pressure control.

$$\text{Torque} = \text{constant} \times \text{speed}^2 \qquad (10.12)$$

Combining this with Eq. 10.8, which says horsepower is directly proportional to the torque, you discover that horsepower in a variable-torque application is proportional to the speed squared, not cubed. The impact of this is dramatic:

Speed (%)	Hp at variable Hp (%)	Hp at variable torque (%)
100	100	100
70	34	49
50	13	25
20	1	4

This implies that all of the assumptions for horsepower savings are overestimated by close to 50 percent. This difference is most exaggerated in the middle of the operating curves, illustrated in Fig. 10.6. This is the area where most of the pump performance occurs and where traditional analyses credit most of the savings. Another way to look at this effect is to imagine its effect on Fig. 10.5. The actual system curve is moved up even further, so that it is nearly horizontal. This takes away most of the little energy savings that were credited to the variable-speed drive.

There are a few applications where VFDs work well with mechanical systems, and they are typically where pressure is not an issue. Cooling tower fans and ventilators can do a substantial part of their work by just moving air against a fairly nominal static pressure. Consequently

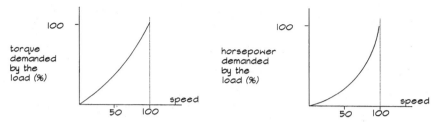

Figure 10.6 Variable torque load.

they are ideally suited to motor speed modulation. Otherwise, the application should be inspected very carefully.

Motor Circuiting

Once the proper size motor has been selected for a specific application, it is a simple matter to power it up. Table 10.3 has all of the important information on three-phase motors—the motor of choice in almost every circumstance. The tabulated values are for induction-type squirrel cage and wound rotor motors only. Notice that the wire size may vary according to the type of insulation used. When specifying a motor for any application, it is best to specify as much information as possible on the plans, including rated efficiency so that when the project is bid and constructed the proper motor is actually installed. (Or there is legal recourse to see that it is installed, at the contractor's expense.)

Table 10.4 contains some more useful information for single-phase ac motors, and Table 10.5 details two-phase ac motors. Information on other motors are available in standard tables.

Some designers always oversize motor feeders by 25 percent, especially in critical applications such as fire pumps, smoke pressurization or exhaust fans, and other safety-related devices. This is usually not a bad practice because erring on the conservative side is OK in any application, as long as it is not taken to the extreme. Doing so will certainly create safe installations, but it will affect profitability and competitiveness. Therefore, it is better to be more knowledgeable in the specific application so that the service can be designed exactly and with confidence. That's another important reason for the electrical designer to develop a solid working knowledge of mechanical systems.

Designing accurately usually means referring to the nameplate data of the exact motor to be installed. However, there is no guarantee that in competitive bidding the selected motor will actually be installed. Therefore, it is best to use conservative values such as in these tables for designing the circuits, feeders, and disconnecting means.

MOTOR STARTER PANEL MSP

BUS: 225 AMPS, 3 PHASE, 3 & GR. WIRE 480 VOLTS
U.L. SHORT CIRCUIT RATING: 30K R.M.S. SYM. AMPS.
MAINS: M.L.O. , 100 AMPS, 3 POLES
MOUNTING SURFACE , NEMA 1 ENCLOSURE

CKT. NO.	SERVES	STARTER SIZE	BREAKERS FRAME AMP	POLE	TRIP	FEEDER NO. & SIZE WIRE	COND.
1	CHP-1		20	3	20	3-#12, 1-#12G	3/4"
2	CHP-2		20	3	20	3-#12, 1-#12G	3/4"
3	CHP-3 & 4 (VFD)		20	3	20	3-#12, 1-#12G	3/4"
4	AHU-1 (VFD)		30	3	30	3-#10, 1-#10G	1"
5	HWP-1		20	3	20	3-#12, 1-#12G	3/4"
6	EF-7		20	3	20	3-#12, 1-#12G	3/4"
7	MAF-1		20	3	20	3-#12, 1-#12G	3/4"
8	SPARES		20	3	20	--	2"
9-11	SPACES		-	-	-	--	-

Figure 10.7 Motor starter panel.

When a specific device is known to be installed, or if applications are checked after the bid is awarded, information that is more detailed should be used. The best source is the manufacturer's nameplate data, which often includes full-load amps and fuse rating of the disconnecting means. Table 10.4 can be used to check these values or to design installations on which little nameplate information is available.

In any installation, the criteria established elsewhere apply. These rules include the 2 percent maximum voltage drop, positive equipment grounding, the proper location of disconnecting means in line of sight and within 50 ft of the equipment, and using waterproof flexible metal conduit for the final connection.

Chapter

11

Transformers

A transformer is a simple device that is used to change the voltage of power in ac circuits. It consists of two sets of windings wrapped around a common iron core. The primary windings are the incoming current, and the secondary windings the outgoing circuit. The ratio of the number of secondary windings to the number of primary windings is the turns ratio. The change in voltage from the primary to the secondary is a direct proportion between the number of windings:

$$\frac{Es}{Ep} = \frac{Ns}{Np} \qquad (11.1)$$

where s = secondary
p = primary
E = voltage
N = number of turns

This proportion is called the voltage ratio. If the primary has 100 turns and the secondary has 1000 turns, the voltage will be stepped up by a factor of 10. A step-down transformer has more turns on the primary than on the secondary and outputs a lower voltage than comes in.

The primary-to-secondary current ratio is also a direct function of the number of turns in the primary and secondary windings:

$$\frac{Ip}{Is} = \frac{Ns}{Np} \qquad (11.2)$$

where I = current

These two relationships signify that the primary power is equal to the secondary power in a transformer that is 100 percent efficient. The efficiency is defined as the ratio of power in to power out:

$$\text{Efficiency} = \frac{P_{in}}{P_{out}} \times 100\% \qquad (11.3)$$

The primary power that is not available in the secondary circuit is dissipated as heat in the transformer. The construction of the core is the primary determinant of the efficiency of the device.

Core Losses

The transformer has no moving parts. Current is the only thing that moves through the primary and secondary coils. As the current moves in the windings around the core, a strong magnetic field is created. This field, surrounding the core, is made up of lines of equal potential that cross the secondary windings. There the magnetic field induces current in the secondary coil. In a perfect transformer all of the magnetic flux lines created by the primary windings cross the secondary windings and result in current being generated there. In less-efficient devices, flux lines stray from the core and dissipate in the surrounding air. They do not induce any current in the secondary windings, so the power used to create them is lost. Closed core transformers, the most common device in building power applications, are an effective design in the reduction of leakage flux from the windings (see Fig. 11.1).

Another potential source of loss is due to resistance in the conductors of the windings. This is a dc resistance called a winding resistance or copper loss. This can be minimized by using larger-size wire for the windings. Sizing the conductors should take into account the rating or application of the transformer. If the transformer is loaded above its capacity, the windings have higher resistance losses. These losses are evidenced as heat, which can damage or reduce the effectiveness of the insulation. This can further reduce the efficiency of the transformer, even at lower current levels. Consequently, winding wire sizes should be increased for applications that operate at a steady load near rated capacity or that frequently peak above that capacity.

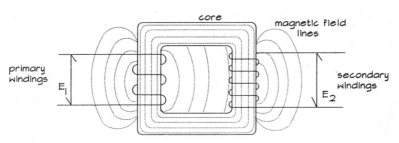

Figure 11.1 The core type transformer.

The core itself is the most significant source of loss. Iron is the most common core material because it has a low reluctance, or resistance to the establishment of magnetic flux lines when a magnetic field is induced. However, iron is also a conductor of electricity. The ac current in the windings induces a current in the core. This current does not do any productive work or benefit the secondary output in any way. It stays in the core and travels endlessly in a small closed path within the core itself. This is called an eddy current, and its outcome is resistance losses (i.e., thermal losses) in the core. The higher the current, the greater the thermal losses and the less efficient the transformer.

These eddy currents, then, are created by the changing magnetic field induced on the core by the windings. In an ac circuit, this field changes 50 or 60 times each second, so the eddy currents are constantly starting, stopping, and changing direction. The higher the frequency of the alternating current, the greater the eddy current losses. This occurs because the eddy current field opposes the coil flux, so more current is needed in the coil to maintain the magnetic field within the core. The eddy currents do no useful work. In fact, all of the power needed to create these short-lived currents is wasted energy. Transformer designs strive to minimize these eddy current losses.

The eddy currents are induced by the encompassing magnetic field because the resistance to current flow in the solid iron core is low. A large core is like a very large-diameter wire; resistance to a small current flow is quite low. If the size of the "conductor" can be reduced and the path lengthened, the resistance to current flow will be greatly increased. Consequently, there will be less opportunity for eddy currents to occur, and when they do, the magnitude of the current will be much lower, so the I^2R losses will be lower. Less magnetic energy will be lost to heat, and the system efficiency will increase.

The eddy current path is increased by laminating the core. Instead of a solid closed core (shaped like a square doughnut), the loop is sliced into many sections, or laminated. As the laminations get smaller, the path length increases (Fig. 11.2) and thus the resistance increases. Consequently, the eddy current flow decreases.

The laminations must be insulated, or the device would still behave like a solid core. The laminations are usually insulated by a thin coating of iron oxide and varnish or lacquer. This layer increases the resistance between laminations to isolate eddy current paths, but it still allows a low reluctance path for the high flux density in the core. Heat can break down the lacquer and cause the effective eddy current paths to increase by reducing the number of effective laminations. This will reduce the efficiency of the transformer.

Another way to improve efficiency is to reduce hysteresis losses, which are caused by the extra power needed to reverse the magnetic

(a) Eddy current in a solid core

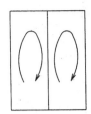
(b) Eddy current paths increased by core lamination

(c) Eddy current paths greatly increased by laminations

Figure 11.2 Action of the transformer core.

field in the core under the influence of the alternating current. The core material is the only factor in hysteresis losses. Once the core construction is designed to minimize eddy currents, the core material can be selected to reduce hysteresis losses even more. Silicon steel has better magnetic performance than iron. The energy needed to magnetize iron with each reverse in polarity is greater than for silicon steel. The effect is the same as if the internal friction between molecules is less, requiring less energy to reorient them with each change in magnetic polarity.

The hysteresis losses can be visualized by inspection of a graph of the magnetic field strength versus the magnetic flux intensity. The system starts at the origin, and as the magnetic field increases, the iron core becomes magnetized. When the material becomes saturated at b in Fig. 11.3, the maximum magnetic strength is achieved. As the field strength is reduced, the core follows the curve through point c, which indicates that some residual magnetism remains in the iron core, making it behave as a permanent magnet. This reverse direction, which is imposed by virtue of the positive and negative sense of an alternating current, continues until point d. When the sense of the magnetic field is reversed again, the lower curve is followed to point b. And the process repeats.

The shape of this curve is significant. It represents the amount of energy that is lost to the core material with each cycle of the current. This

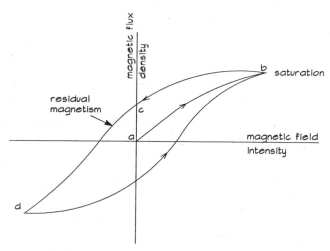

Figure 11.3 The magnetic hysteresis loop.

occurs 60 times a second for 60-cycle power. The laminations and other modifications to the solid iron core reduce the area enclosed by the hysteresis loop, thus reducing the energy that is dissipated in the device.

Later the discussion will turn to the harmful effect of harmonics to a power system. Harmonics are smaller multiples of the driving power. For example, the common third harmonic is 3 times the frequency of the fundamental frequency, or 180 cycles per second. This waveform, then, will cause the hysteresis loop to be repeated much more rapidly; the common fifth harmonic will excite the core at 300 Hz. This dissipates a great deal of energy in any iron core, either in a transformer or a motor.

Applications

Although they should not be used indiscriminately in a power distribution system, transformers do have their uses. Foremost among them is that they allow power to be distributed at a higher voltage. This means smaller conductors, smaller conduits, and less material and labor for installation. It also means that higher-voltage power can be available for special applications throughout a facility. For example, it is usually best to use 208- or 240-V motors for anything other than fractional horsepower motors and 480-V three-phase motors for applications greater than 5 or 10 hp. It is also more economical to have 277 V available for fluorescent lighting circuits. It is important to try to accommodate these needs if it is compatible with the electrical layout.

As facilities get larger, they often have a separate equipment circuit for large fan motors, air handler motors, and other electromagnetic equipment. Distributing the power to each high-voltage panel and stepping it down to feed the panel is often very economical, especially if the current needs are high and busways are doubled-up.

A transformer is just a coil when it has no load on the secondary side. Only when a load is applied to the output side does the device perform as a transformer. Therefore, whenever there is any load, current flows in the primary and secondary coils. Eddy currents are developed then, and hysteresis losses reduce the efficiency of the system and cause useful energy to be discharged as heat. Energy use can be reduced if the circuiting is configured so that no power is used by the secondary at all during off-hours. This reduces electricity use and the hvac load because the transformer is not emitting any heat into the conditioned space.

One practical measure of the transformer efficiency is called the all-day efficiency. This is defined as

$$\text{All-day efficiency} = \frac{W_{in}}{W_{out}} \times 100\% \qquad (11.4)$$

where W_{in} = total energy delivered during 24 h
W_{out} = total energy supplied during 24 h

This criteria is important in ac systems when the transformer is connected to the building load for the whole period. The core losses persist during the whole time, and the copper losses vary according to the load. The design of a transformer for this type of application is different from the device used in applications when the load is disconnected from the transformer during off-hours. Full-time transformers are designed with a lower core loss (i.e., more laminations) and a higher copper loss (i.e., smaller diameter windings).

Three-Phase Connections and Polarity

The study of transformers as single-phase devices is quite straightforward. The secondary windings can be split to provide, for example, 120- and 240-V power from the same primary service. The primary voltage may not always be at the specified level, so several taps are provided on the primary windings. As illustrated in Fig. 11.4, the taps vary the number of turns on the primary side. They can be adjusted to fix the secondary voltage at the design value.

Standard transformers can be operated at a voltage output of not more than 10 percent above design voltage. Higher voltages overheat the coil, which can damage the insulation of the core and windings. Lower voltages do not cause overheating if the rated current is not ex-

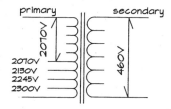

Figure 11.4 Transformer taps.

ceeded. However, the power rating is reduced. As a result, it is always important to use the taps to adjust the primary voltage to the design value.

Three-phase systems are much more complex than single-phase systems. There are several ways to configure the taps on the secondary windings to provide different voltages and loading capacities. Another factor involves induced voltages. Voltage applied to the primary coil magnetizes the coil and creates a strong magnetic field in the secondary windings, inducing voltage. This voltage in the secondary coil opposes the voltage in the primary coil. This countervoltage creates voltage in the second coil by induction.

This effect must be taken into account when connecting several transformers in a system. There are two possibilities, additive and subtractive polarity. All of the secondaries and primaries on the transformers should have the same polarity. Figure 11.5 illustrates the two schemes. The diagrams use the standard notation for labeling terminals of transistors, with the letter H subscripted with numbers at the primary windings. The low-voltage connections on the secondary windings are identified by the letter X with subscripted numbers. If the windings are connected in additive polarity, the induced voltage in the primary windings will be opposite the induced voltage in the secondary windings. If the polarity is subtractive, the primary and secondary windings will be connected in the same direction. If the windings are not connected as shown, the transformer will malfunction. The fault will either trip the overcurrent protection device, or the transformer will operate improperly.

Figure 11.6 shows several additional connecting means between the different kinds of phased circuits. The specific configuration of each three-phase connection is discussed in the next section.

Delta connections

This system is usually represented by a simple triangle. Only one voltage is available between any two wires in a delta system. The delta connection is good for short-distance systems, such as smaller commercial loads and neighborhood loads that are relatively close to the electrical substation.

176   Chapter Eleven

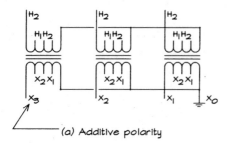

(a) Additive polarity

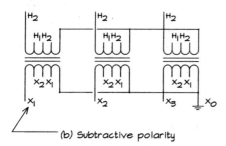

(b) Subtractive polarity

Figure 11.5 Connecting three-phase transformers.

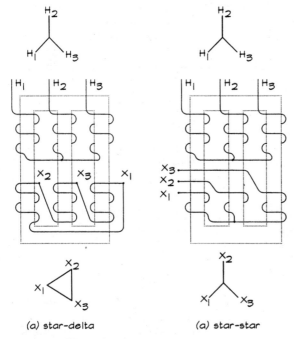

(a) star-delta   (a) star-star

Figure 11.6 Three-phase transformer connections.

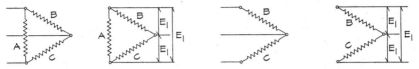

Figure 11.7 Open delta connections.

One of the big advantages of a delta system is that if one of the transformers is removed from the circuit, the remaining transformers can still maintain the correct voltage and phase relations in the secondary (see Fig. 11.7). The load must be reduced accordingly, to not exceed 86 percent of the rating of the two remaining transformers. This is another reason the delta system is used to supply power to individual customers.

Wye or star connection

The wye system has three hot conductors with one neutral. This means that 120 V is available from the three-phase conductors throughout the building (Fig. 11.8). This is a distinct advantage for commercial buildings because of the flexibility of having three-phase power available, plus 120 V at any point of use.

If one of the transformers develops trouble in a wye-connected system, it is not possible to operate the remaining two transformers to continue service, even at a diminished level. Service cannot be renewed until the faulty transformer has been replaced or repaired.

One advantage of a wye-connected system is that each transformer must be designed for 60 percent of the line voltage. Transformers in a delta system must be sized for full line voltage. In high-voltage systems there is considerable savings in not having to insulate the transformer for full voltage, so wye connections are preferred on the high-voltage primary side. The low-voltage side is done in delta for dependability in the event of a transformer fault.

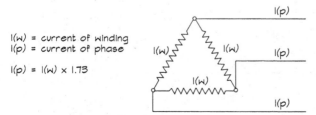

Figure 11.8 Three-phase delta connection.

Wye-delta connections

This application is best for large installations that have a large power demand. The best use is in industrial applications where a large amount of three-phase current is required, but only a small amount of single-phase 120-V load.

Isolation Transformers

Normally, because one side of the ac power line is grounded, connecting grounded equipment to the grounded line can cause a short circuit. The only way in conventional circuits to prevent this is to ensure the grounded side of the equipment is connected to the grounded side of the power line. Equipment requiring such protection is often used at several locations, and power outlets in the building—or even in the same room—are not usually polarized. Consequently, it is practical to use an isolation transformer. The chassis becomes a ground, and it is virtually impossible to have a short circuit under such conditions.

Isolation transformers are used in hospital operating rooms and emergency rooms so that there is not any danger of short circuiting equipment at a critical time or the threat of shock to patients and staff. The isolated ground also ensures accurate operation of very sensitive and important electronic equipment. Any business that operates sensitive electronic equipment can benefit from isolated systems as well. See Fig. 11.9 for a schedule of transformer ratings.

TRANSFORMER SCHEDULE

MARK	DESCRIPTION	PRIMARY	SECONDARY
TA	30 KVA, 3 PHASE, DRYTYPE	480 V THREE PHASE	120/208 VOLT, THREE PHASE, FOUR WIRE
TB	30 KVA, 3 PHASE, DRYTYPE	480 V, THREE PHASE	120/208 VOLT, THREE PHASE, FOUR WIRE
TC	30 KVA, 3 PHASE, DRYTYPE	480 V, THREE PHASE	120/208 VOLT, THREE PHASE, FOUR WIRE
TK	45 KVA, 3 PHASE, DRYTYPE	480 V, THREE PHASE	120/208 VOLT, THREE PHASE, FOUR WIRE
TE1	15 KVA, 3 PHASE, DRYTYPE	480 V, THREE PHASE	120/208 VOLT, THREE PHASE, FOUR WIRE
TE2	30 KVA, 3 PHASE, DRYTYPE	480 V, THREE PHASE	120/208 VOLT, THREE PHASE, FOUR WIRE
TE3	45 KVA, 3 PHASE, DRYTYPE	480 V, THREE PHASE	120/208 VOLT, THREE PHASE, FOUR WIRE

Figure 11.9 Transformer schedule.

Chapter 12

Estimating

Despite everything that has been said, sometimes there is just no better way to learn something than by doing it. The next best thing to doing it is to estimate it. Every home run must be routed just the way it will actually run, every junction box located and every switch wired, circuit grounded, and fixture installed, wired, and switched. By doing this simple exercise, the designer will become more cognizant of how useful his or her finished plans are to the contractors bidding the work. The best plans can be taken off swiftly and with confidence. This means the average bid prices will be lower and more contractors will bid on the work, both of which bode well for the success of the project. Some of the most important things that make this possible are

1. Hot wire and switch legs are shown for lighting, plus all the junction boxes needed to assemble the system are shown. This is not much more trouble to do with computer drafting tools than using tick marks and schematic circuits, and it is a lot more helpful for the contractors.

2. Panelboard schedules are complete, accurate, and reasonably balanced between phases. The better the designer can manage to balance the load per phase, the closer the contractor can then refine it in the field.

3. Load, voltage, and full-load amps needed at the point of use are identified for all nonstandard equipment loads.

4. Every segment of wire shown anywhere on the plans is clearly labeled with line size, ground wire, and conduit size (if a feeder is shown in two places on the plans, only size it at one location, whichever is more schematic, such as the riser).

5. Light fixtures are located accurately in the ceiling grid, switches are out of door swings, and receptacles are accessible.

6. All circuits are drawn as they will be installed, with lines drawn parallel to building elements, points of use in sequence (series), and home runs pointing from the closest point of the circuit to its panel.

These are all the things that, if done improperly or incompletely, will cause the estimator to increase the contingency on the job above the nominal.

Service Corridors

The single most helpful thing that a designer can do for the contractor is to negotiate with the other design team members a specific portion of the ceiling space to carry electrical service. This is especially important above corridors, where hvac, plumbing, electrical, and fire alarm designers all like to run their main lines. Usually the two space hogs, hvac and plumbing, are drawn schematically on the plans leaving the contractors to negotiate space allocations and conflicts in the field (and then leaving the electrical contractor what odd configurations of space that remain). A little planning and cooperation will make all this redesign unnecessary if the disciplines stay in their designated areas.

Once the drawings are marked up sufficiently to at least resemble the actual shop drawings, it is time to start estimating. Fixtures, receptacles, junction boxes, and all other items are tallied up, and cost is calculated for materials. The same is done for wiring, conduit, panels, disconnects, and all the other accoutrements needed for a complete job. Then comes the most difficult task of all: estimating labor costs. The tables referenced in the rest of this section give a good idea of the time involved for the major tasks.

As the materials and labor charts are slowly assembled, the deficiencies of the plans become more obvious and more frustrating. Not only are items missing, incorrect, and drawn wrong, but other aspects may actually have to be redesigned to comply with the code. The estimator or contractor is not supposed to have to do any design work. It is hard enough to assemble the estimate in the first place. When the electrical engineer's job is not done well, the system efficiency decreases, and the project cost escalates.

The only tools needed to take off an electrical job are a scaled ruler, a highlighter, and some lined paper with columns such as accountants use. As each item is counted, it is marked out and tallied on the paper. In the following synopsis the work is assumed to be done by a contractor who will be bidding the project. The same method should be followed by an electrical engineer taking off the work for a preliminary budget cost for the owner prior to bid time. In either case it is a good idea to keep a separate pad available to jot down questionable items, either from the design, cost, or installation point of view.

TABLE 12.1 Lighting Fixture Labor Values*

Length of fixture	Number of lamps	Description	Mounting method		
			Ceiling	Recessed	Pendant
4	1	Strip, bare	1.24	1.75	1.55
4	2	Strip, bare	1.39	1.89	1.69
4	4	Strip, bare	2.06	2.16	2.36
4	2	Strip, w/lens	1.74	2.25	2.05
4	4	Strip, w/lens	2.56	2.66	2.86
4	2	Troffer	2.10	2.60	
4	4	Troffer w/lens	3.05	3.16	
	Hi Hat		1.20	1.49	1.70

*Hours per fixture. Labor includes mounting, fastening, connecting, and lamping. Essentially "ready to use."

Lighting Fixtures

The first task is to count light fixtures by fixture type. (Refer to Table 12.1 for labor values.) The preferred method is to mark off each fixture as it is counted. Doing one sheet at a time, the total of the fixture type on that sheet is written on the tally sheet before going to the next sheet. This prevents a serious loss of data if there is an interruption. Counting one fixture type all through the plans means the estimator goes through the entire set several times, as each fixture type is totaled. With each pass, fixtures may be noticed that were missed, that might not have been noticed if all fixtures were taken off each page at the same time. Going through the entire set several times also gives the estimator a good feel for the project and the floor plan layout before starting to take off the more hard-to-find items from the plans.

Devices

The first device to take off is the switching for lights and outlets, as applicable. (Refer to Table 12.2 for labor values.) The one-, three-, and four-way devices should be counted on each sheet, not to include motor disconnects and motor-rated switches. This completes the take off of the lighting plans. Note that the conduit and wire take off is not done yet.

The next devices to consider are the receptacles on the power plans. First, all the duplex receptacles are counted, then the single high-voltage and heavy-duty power devices are counted.

The miscellaneous system devices are counted next. They are sometimes located on the power drawings, but for complex systems are situated on a separate series of drawings. The devices in this category are

TABLE 12.2 Device Labor Values*

	Ampere rating			
	15 (#14 wire)	20 (#12 wire)	30 (#10 wire)	50 (#6 wire)
Single-pole switches S1	19.00	24.38	31.67	
Double-pole switches S2	39.40	51.00	61.67	
Three-way switches S3	30.40	35.67	46.66	
Four-way switches S4	39.40	51.00	61.67	
Duplex receptacles	21.30	25.38		
Duplex receptacles, 2 cir.	31.50	34.00		
Single receptacles, 115/250 V, three pole	30.17	39.88	50.70	66.25
Clock receptacle	32.50			92.50†
Fan receptacle	51.50			
Plates, per gang	8.66			
Weatherproof plates	12.50			

*Units figured at 100 pieces per hour.
†60 A.

- Smoke detectors, ceiling and duct mounted
- Heat sensors
- Fire alarm devices including pull stations, flow sensors, and signaling devices
- Security devices such as motion detectors and glass break sensors

Some of the devices, such as duct-mounted smoke detectors, may be on the mechanical plans. If they are not shown on the electrical sheets, the hvac plans will have to be consulted.

As with any section, once the entire take off is complete, it is a good idea to scan back through the plans again to check for any items that may have been missed. If the first count was done horizontally from top to bottom, it may be beneficial to take a different approach for the review, scanning from right to left instead.

Motors

There are usually not many motors on a project. (Refer to Table 12.3 for motor labor values.) Most motors are priced as part of an assembly, such as a fan, pump, air handler, or elevator motor. Usually the specifications say who is responsible for these motors—the hvac, plumbing, or electrical contractor. It is still important to specify each motor, noting FBO (furnished by others) in the table for those that are not furnished by the electrical contractor, because the disconnect and feeder are still the responsibility of the electrical contractor. The applicable

TABLE 12.3 Unit or Modular Motor Labor Values*

hp	Poles	Volts	Wire size	Starter size	Labor	Disc. size	Switch labor	Conn. labor	Total labor
Fract.	2	230	14		1.99	Tgle		.43	2.42
Fract.	3	230	14		2.20	Tgle		.65	2.85
Fract.	2	460	14		2.36	Tgle		.43	2.79
Fract.	3	460	14		2.72	Tgle		.65	3.37
1	2	230	14	00	2.52	30	2.17	.43	5.12
1.5–2	3	230	14	00	2.82	30	2.17	.65	5.64
1.5–2	3	460	14	00	2.82	30	2.45	.65	5.92
1–2	2	230	14	0	2.52	30	2.17	.43	5.12
3	3	230	14	0	2.82	30	2.33	.65	5.80
3	2	230	14	1	2.60	30	2.17	.43	5.20
5	3	460	14	0	2.82	30	2.47	.65	5.94
7.5	2	230	6	2	3.27	60	2.50	.97	6.74
7.5	3	230	10	1	3.12	60	2.96	.97	7.05
10	3	460	12	1	2.94	30	3.65	.83	7.42
15	3	230	6	2	3.52	60	3.33	1.46	8.31
25	3	460	6	2	3.52	60	4.35	1.46	9.33
30	3	230	1	3	4.49	100	4.23	2.20	10.92
30	3	460	6	3	3.52	60	4.78	1.46	9.76
40	3	230	4	3	4.07	100	5.26	1.62	10.95
40	3	460	2/0	4	4.97	200	6.36	3.73	15.06
50	3	230	3/0	4	5.27	200	7.00	4.21	16.48
50	3	460	2	3	4.37	100	5.79	1.78	11.94
60	3	230	4/0	5	7.22	200	7.70	4.86	19.78
60	3	460	1	4	4.49	100	6.37	2.20	13.06
75	3	230	300 MCM	5	7.82	400	10.53	6.38	24.73
75	3	460	1/0	4	4.67	200	8.20	3.24	16.11
100	3	230	500 MCM	5	8.72	400	11.58	8.26	28.56
100	3	460	3/0	4	5.27	200	9.02	4.21	18.50

Devices Mounted in Starter, Not Wired	
Stop or start station	.30
Stop and start station	.45
Stop and start station with pilot light	.75
Remote stop or start station in NEMA 1 enclosure	1.30
Remote stop and start station in NEMA 1 enclosure	1.45
Remote stop and start station in NEMA 1 enclosure with pilot light	2.00

*Labor values in hours. Essentially "ready to use."

motors are tabulated by important characteristics such as horsepower, voltage, phase, efficiency rating, and casing type or application.

Starters are the next thing to count. (Refer to Table 12.4 for starter labor values.) Again, they may or may not be supplied by others, but they are certain to be wired under the electrical contract, so it is best to count all starters. They are tallied by size, voltage, and any other important characteristics such as soft starting, hand-off-auto, pilot lights, limit switches, alarms, or control interlocks. Many installations are using variable-speed drives for motors of all types. These too must be

TABLE 12.4 Magnetic Across-the-Line Starters*

Cont. current rating	Starter size	TW wire size, CU†	Max. Hp	Devices Two Conn. 230 V	Three Conn. 230 V	Three Conn. 480 V
9	00	14	Frct-1	2.52		
9	00	14	1.5–2		2.82	2.82
18	0	14	1–2	2.52		
18	0	14	3		2.82	
18	0	14	5			2.82
27	1	12	3	2.60		
27	1	10	7.5		3.12	
27	1	12	10			2.94
45	2	6	7.5	3.27		
45	2	6	15		3.52	
45	2	6	25			3.52
90	3	1	30		4.49	
90	3	6	30			3.52
90	3	4	40			4.07
135	4	2/0	40		4.97	
135	4	3/0	50		5.27	
90	3	2	50			4.37
270	5	4/0	60		7.22	
135	4	1	60			4.49
270	5	300 MCM	75		7.82	
135	4	1/0	75			4.67
270	5	500 MCM	100		8.72	
135	4	3/0	100			5.27
540	6	750 MCM	125		16.86	
270	5	4/0	125			7.22
540	6	1000 MCM	150		17.16	
270	5	300 MCM	150			7.82
540	6	2-500 MCM	200		21.36	
270	5	500 MCM	200			8.72

Mounted in Starter, Not Wired	
Stop or start station	.30
Stop and start station	.45
Stop and start station with pilot light	.75
Remote stop or start station	1.30
Remote stop and start station	1.45
Remote stop and start station with pilot light	2.00

*Hours per unit. Labor includes stripping, connecting, drilling, mounting, fastening NEMA 1 enclosures. Line and load connections and two heaters are included.
†Wire size computed at 125% of full load of motor.

totaled. Finally, depending on the number of motors and their location, it may be economical to install a motor control center.

If the quantity of motors, starters, and drives is considerable, it may be advisable to submit the final tally to an equipment supplier for a quick budgetary quote. A supplier who will be bidding on the job anyway will not mind helping and may even have some good ideas on alternate selections and specifications. Soliciting outside quotes on all major items is a good idea because they are invariably lower than the equipment estimating manuals provide. An alternative is to use current pricing sheets provided by major manufacturers, with their current price multiples for the list costs. (Refer to Table 12.5 for motor connection labor; Table 12.6 for transformer labor; and Tables 12.7 and 12.8 for disconnect labor.)

The remaining devices are relatively easy to take off the plans.

TABLE 12.5 Connections at Motors*

Max. hp motor	TW Insul. wire size, CU	Two-pole 230 V	Three-pole 230 V	Three-pole 460 V
Fract-1	14	.43		
1.5–2	14		.65	.65
1–2	14	.43		
3	14		.65	
5	14			.65
3	12	.54		
7.5	10		.97	
10	12			.83
7.5	6	.97		
15	6		1.46	
25	6			1.46
30	1		2.20	
30	6			1.46
40	4			1.62
40	2/0		3.73	
50	3/0		4.21	
50	2			1.78
60	4/0		4.86	
60	1			2.20
75	300 MCM		6.38	
75	1/0			3.24
100	500 MCM		8.26	
100	3/0			4.21
125	750 MCM		12.96	
125	4/0			4.86
150	1000 MCM		15.00	
150	300 MCM			6.38
200	2-500 MCM		16.72	
200	500 MCM			8.26

*Hours per unit. Labor includes stripping, connecting, taping, either split connection or lugs in motor terminal box. Motor is mounted in place by others.

TABLE 12.6 Dry-Type Transformers: 480-240 Primary, 240-120 Secondary*

		Primary				Secondary				Mounting method			
		480		240		240		120					
kVA	Ø	W	A	W	A	W	A	W	A	1Ø floor	1Ø wall	3Ø floor	3Ø wall
.50	1	14	2			14	4	14	6	2.93	2.93		
.50	1			14	3			14	6	2.93	2.93		
.75	1	14	2			14	4	14	7	3.18	3.18		
.75	1			14	4			14	7	3.18	3.18		
1	1	14	3			14	6	14	9	3.43	3.81		
1	1			14	5			14	9	3.43	3.81		
1.5	1	14	4			14	7	14	13	3.68	4.15		
1.5	1			14	7			14	13	3.68	4.15		
2	1	14	5			14	9	12	17	3.93	4.62		
2	1			14	9			12	17	3.98	4.67		
3	1	14	7			14	13	10	25	4.21	5.09		
3	1			14	13			10	25	4.48	5.35		
3	3	14	4			14	8					4.93	5.93
5	1	14	11			14	21	6	42	5.27	6.60		
5	1			10	21			6	42	5.54	6.88		
5	3	14	8			14	15					6.43	7.73
6	3	12	16			8	32	4	63	6.63	7.69		
7.5	1	14	11			10	22	4	63	7.42	8.48		
7.5	1			8	32			4	63	8.50	9.87		
9	3	14	11			6	42			8.87	10.58		
10	1	10	21			10	42	2	84	9.42	11.13		
10	1			6	42			2	84	10.96	14.32		
15	3	8	32			4	63			12.36	15.72		
15	1			4	63			1/0	125	14.54	19.96		
15	1	12	19			8	37	1/0	125	17.42	20.57		
25	3	6	53			1	105					10.40	12.05
25	1	8	37			3	74	250	209	19.91	25.65		
30	3			1	105			250	209			12.23	14.60
37.5	1	3	79			3/0	158	3/0	316	24.19	29.93		
37.5	1			3/0	158			3/0	316				

kVA	Ø	W	A	W	A	W	A	W	A	$	$	$	$
45	3	6	55									19.07	21.94
50	1	1	105			1	110						
50	1	3/0	157	250	210	250	210			24.02	30.13		
75	1			500	314	500	314	250	420	30.25	36.36		
75	1							500	628	30.02	38.82		
75	3	2	91			4/0	182			38.23	47.03	26.69	29.94
100	1	250	209	250	418	250	418			38.05	43.75		
100	1							400	836	53.10	58.80		
112.5	3	2/0	136			400	272					37.43	43.93
150	3	4/0	181			4/0	362					43.22	49.72
167	1												
167	1	600	348	3/300	698	3/300	698	5/400	1396	70.44	76.74		
225	3	400	271			2/400	542			88.28	94.58	65.85	81.85
300	3	700	362			3/350	724					83.22	102.32

*Hours per unit. kVA = kilovoltamperers; W = conductor size; A = ampere; Ø = phase; Labor includes mounting, fastening, wire connection (split bolt or lug), and taping. Essentially "ready to use."

TABLE 12.7 Motor and Disconnect Switches, Externally Operable (250 V)*

Hp fractional	Switch size, amperes, at 250 V, toggle, motor rated	Two connections	Three connections	Four connections
		1.99	2.20	2.31
1–3	30	2.17	2.33	2.45
5	30	2.24	2.47	2.60
7.5	60	2.50	2.96	3.34
10	60	2.65	3.14	3.54
15	60	2.81	3.33	3.75
20	100	3.16	3.76	4.50
25	100	3.35	3.99	4.77
30	100	3.55	4.23	5.06
40	200	5.43	6.36	7.31
50	200	5.98	7.00	8.04
60	200	6.58	7.70	8.84
75	400	8.68	10.53	12.47
100	400	9.55	11.58	13.72

*Hours per unit. Labor includes securing, mounting, fastening, connecting, and identifying. Essentially "ready for use."

TABLE 12.8 Motor and Disconnect Switches, Externally Operable (600 V)*

Hp fractional	Switch size, amperes, at 600 V, toggle, motor rated	Two connections	Three connections	Four connections
		2.36	2.72	3.21
1–3	30	2.50	2.88	3.40
5	30	2.65	3.05	3.61
7.5	30	2.82	3.23	3.83
10	30	3.10	3.65	4.25
15	30	3.29	3.87	4.50
20	60	3.49	4.10	4.78
25	60	3.70	4.35	5.07
30	60	3.92	4.78	5.37
40	100	4.31	5.26	6.11
50	100	4.74	5.79	6.72
60	100	5.22	6.37	7.39
75	200	6.65	8.20	9.22
100	200	7.13	9.02	10.14

*Hours per unit. Labor includes securing, mounting, fastening, connecting, and identifying. Essentially "ready for use."

TABLE 12.9 Conduit Labor Values*

Size (in)	Rigid galvanized HW	Aluminum rigid HW	Electrical metallic tubing	Polyvinyl chloride
1/2	4.75	4.75	4.00	4.50
3/4	5.65	5.90	4.75	5.50
1	7.45	7.25	6.40	6.75
1 1/4	8.70	7.45	7.10	7.00
1 1/2	10.30	8.85	8.00	7.25
2	13.50	11.00	9.70	10.25
2 1/2	18.45	13.90	11.85	12.75
3	22.55	17.00	14.00	15.00
3 1/2	27.25	20.40	17.00	18.00
4	35.70	25.15	18.00	24.50
5	50.10	36.70		33.00
6	65.00	49.00		44.00

*Hours for 100 ft of installation. Includes fastenings, mounting, and terminating.

Conduit Costs and Labor

The methods established earlier in the text for designing power systems describe fully the means to route home runs by the most expeditious route. (Refer to Table 12.9 for conduit labor values and Tables 12.10 to 12.12 for the conduit-related labor values.) Two or three home runs can sometimes be combined into a single conduit, although a conservative estimate is to just assume all home runs are in dedicated conduits. The savings from combining home runs will be offset by extra costs due to conflicts with other disciplines, space requirements, and other such practical concerns. (Refer to Tables 12.13 and 12.14 for panel labor values.)

Additional Factors

Once the material and labor costs have been generated, there are a number of additional costs that need to be included in the pricing. These are as follows:

1. Sales tax, shipping, and storage of materials.
2. Labor productivity factor to account for the skill level of the field personnel.
3. Nonproductive labor cost for staff support individuals, such as stockers, clerks, drivers, drafters, supervisors.

TABLE 12.10 Wire Pulling Labor Values*

Size of wire (gauge)	Copper		Aluminum	
	Solid	Stranded	Solid	Stranded
16	5.00	6.00		
14	6.25	6.90		
12	7.65	8.00	6.25	7.25
10	9.00	10.40	9.00	10.00
8		11.95		11.00
6		13.05		12.00
4		16.40		14.00
3		18.00		
2		18.70		15.25
1		19.55		16.25
1/0		24.60		20.00
2/0		27.65		21.25
3/0		30.00		24.25
4/0		36.70		27.75
250 MCM		37.70		28.75
300 MCM		42.10		30.75
350 MCM		43.50		32.75
400 MCM		44.00		34.75
500 MCM		50.00		36.00
600 MCM		57.00		41.50
750 MCM		64.00		46.00
1000 MCM		78.00		55.50

*Hours per 1000 ft of installation. All insulation is valued using thermoplastic-type insulation. No. 16-10 are computed as "branch circuits," less length and more 90s. No. 8-1000 MCM evaluated as feeder and figured on an average of 150-ft length per run. All wires are left hanging in their panels or boxes.

4. Supervision in the field to include staff progress meetings with the architect, engineer, or construction manager.
5. Travel time to the job site, for productive as well as nonproductive individuals.
6. Expenses such as field radios, duplication of plans, safety equipment.
7. Special safety training, such as required for trenching and scaffolding by OSHA.
8. Temporary light and power installation and power bill payment.
9. Temporary wiring for light and power.
10. Lost time because of accidents, coordination with other disciplines, and design conflicts with the engineer or architect.
11. Overtime.
12. Liquidated damages if such are written into the contract.
13. Permit and inspection fees, including aid to construction fees.

TABLE 12.11 Conduit and Wire Labor Values (Labor Values Combined into One Value)*

Conduit size/ gauge of wire (in)	Electric metallic tubing		Rigid galvanized conduit	
	Three-wire	Four-wire	Three-wire	Four-wire
1/2/14	6.28	7.03	7.02	7.79
1/2/12	6.64	7.52	7.39	8.27
1/2/10	7.43	8.58	8.18	9.33
1/2/8	7.94		8.69	
3/4/8		10.01		10.91
1/6	10.71	12.14	11.75	13.20
1/4	11.81		12.86	
1 1/4/4		14.32		15.92
1 1/4/3	13.04	15.02	14.64	16.62
1 1/4/2	13.27	15.33	14.87	16.93
1 1/4/1	13.55	16.60	15.15	18.90
1 1/2/0	16.12		18.42	
1 1/2/00	17.12		19.42	
2/0		20.52		24.32
2/00		21.87		25.67
2/000	19.60	22.90	23.40	26.70
2/0000	21.81		25.61	
2 1/2/0000		28.00		34.60
2 1/2/250 MCM	23.99	28.14	30.89	35.04
2 1/2/300 MCM	25.44		32.34	
3/300 MCM		32.52		41.07
3/350 MCM		33.14		41.69
3/400 MCM	28.52	33.36	37.07	41.91
3/500 MCM	30.50		39.05	
3/600 MCM	32.81		41.36	
3 1/2/500 MCM		39.00		49.25
3 1/2/600 MCM		42.08		52.33
3 1/2/750 MCM	38.12		48.37	
4/750 MCM		46.16		63.86

*Units are based on 100 ft of conduit, the number of wires in the conduit times 100, and 10 percent for typing in, etc.

14. Security, parking.
15. Telephone and mobile phone charges.
16. Rigging, such as required to hoist motors or switchgear to elevated locations.
17. Direct job expenses for renting or purchasing equipment needed specifically for the project (purchased equipment costs can be amortized over several years and projects)

TABLE 12.12 Armored and Nonmetallic Cable Labor Values*

Gauge of conductors	Armored cable			Nonmetallic cable	
	Two-wire	Three-wire	Four-wire	Two-wire	Three-wire
14	2.49	2.83	3.50	2.36	2.73
12	2.77	3.17	4.00	2.73	3.12
10	3.78	4.13	5.00	3.62	4.38
8	4.94	5.19	7.50	4.60	5.57
6	5.30	5.50	8.80	6.13	
4	6.40	6.75	10.60		
2	6.80	7.00	11.20		
1	7.90	8.25	13.00		
1/0	8.30	8.50	13.80		
2/0	9.40	9.75	15.60		
3/0	10.00	10.25	15.90		
4/0	11.50	11.75	18.70		
250 MCM	12.00	12.50	20.00		
350 MCM	13.00	13.75	21.80		
500 MCM	14.00	15.50	23.80		
750 MCM	16.50	18.00	27.50		

*Hours per 100 ft of cable.

TABLE 12.13 Panel Labor Values*

Panel size (A)	Two poles	Three poles	Four poles
30	2.45	2.57	2.85
100	3.58	4.06	5.02
225	4.70	6.82	7.91
400	9.18	10.98	13.13
600	12.28	15.10	17.32
800	18.63	21.29	24.60
1000	20.82	23.65	25.80
1200	23.00	27.00	29.00

*Labor values include unpacking, removing the trim, removing the interior, storing both trim and interior and mounting the back box in place. Fastenings are included.

18. Testing of equipment such as an emergency generator, to include supplying a standby person or factory expert, extra fuel, down time, and special equipment costs.
19. Productivity factor to apply to the original labor rates.
20. Complexity allowance for coordination and other unknowns. This is especially important if a project is a remodel done in phases, if the facility will continue to be in use during construction, and other such considerations.
21. Weather days.

TABLE 12.14 Connections to Lugs in Panels and Switches*

Size of breakers	Size of conductor with T-TW insulation	Two connections	Three connections	Four connections
15–40	14–8	.65	.96	1.26
55–95	6–2	.85	1.46	1.89
110	1	1.43	2.07	2.60
125	1/0	1.73	2.39	3.03
145	2/0	1.93	2.62	3.32
165	3/0	2.13	2.86	3.65
195	4/0	2.32	3.09	3.94
215	250 MCM	2.58	3.43	4.42
240	300 MCM	2.82	3.85	4.95
260	350 MCM	3.05	4.24	5.47
280	400 MCM	3.33	4.62	6.01
320	500 MCM	3.64	5.01	6.61
355	600 MCM	4.04	5.51	7.28
400	750 MCM	4.65	6.06	8.03
455	1000 MCM	5.62	7.25	9.03

*Hours per unit. Labor includes replacing interior, trim, and directory, plus lacing and stripping conductor and connecting.

22. Furnishing of as-built drawings, especially if on CAD. The contract should clearly specify that the owner will provide the architectural background drawings in CAD format, plus the original electrical design drawings, if the as-built drawings are to be electronic.
23. Escalation of prices for a long job or one that can potentially be delayed, such as a government job.
24. Bid and performance bond costs.
25. Insurance.
26. Overhead such as office rent, telephone, taxes, salaries, advertising, accounting, and other professional services.
27. Profit.

This list is a brief tally of all of the factors that go into the cost of a job. Designers and engineers rarely take all of them into account, resulting in a low estimated budget. This can cause grievous problems when the bids come in, especially for design-build projects. The owner is usually quite upset when bids come in far above the estimates and not so much because of the extra money the project will cost. The problem is that the owner's entire administrative budget process will have to be repeated to solicit the extra funds from the board or other governing authority. Having to do so makes the administrator, in turn, look bad—and the board too—before members of the organization at large because the project designer was not diligent in taking off the project. It is far, far better to overestimate the funds initially and to fight all of those battles only once.

Another, equally important, reason to become intimately familiar with the cost estimating process is that it allows the designer to provide all of the necessary information, clearly and accurately, so that the contractors can manage an accurate take off of the project plans and specifications. Again, contractors are not too worried about the presence of expensive, complex equipment on a project as long as it is necessary and it is fully specified and described. What really bothers them is what is not clearly shown on the plans or not shown at all. If something unusual is buried in the specifications, for example, it leaves the impression that the designer is not being honest and straightforward, which may or may not be the case. When something like that happens, the contractor usually will add a sizable contingency cost to the project bid to cover other such hidden costs in the small print. Consequently, it usually results in lower bid costs to not only put all of the important information on the plans (or to refer from the plans to explicit, unusual sections of the specifications) but to also completely describe all of the important facets of the installation. It usually costs more to try to hide important criteria than not doing so. In fact, it is so rare for design engineers to be so completely honest that if one is, contractors will reduce other contingencies in the bid, anticipating a speedy and productive project with few misunderstandings. The result is an even lower bid price still.

Chapter

13

Computer Applications

There are many software programs on the market, designed to perform many of the calculations required of electrical engineering designers. The most common, and useful, applications are

- Voltage drop
- Zonal cavity lighting
- Short circuit analysis
- Generator sizing

Other programs are often used in the design process by specialists such as manufacturer's representatives because they are more complex and expensive. These include

- Point-by-point lighting
- Coordination analysis
- Daylighting analysis

It is not uncommon to encounter a large engineering firm that does not use any of these calculation methods. The analyses are either done manually, or the sales representatives as a courtesy do them so their products can be specified. It is an irony that, of all the engineering disciplines, electrical is the least computerized.

There are many quality computer programs on the market. Typically they are far more complicated than the typical designer needs, however. As an example, consider the lighting analysis software on the market. The designer really only needs a handful of the most common fixtures as a selection set, plus a couple of possible replacement lamps. A program with thousands of fixtures and hundreds of bulbs is not a useful tool. This huge selection and, indeed, the majority of the fancy features and capabilities of the lighting design software are not only not needed, but also unwanted. They clutter up the program, making it slower to run, harder to learn, and more cumbersome to use. It is often easier to do the analysis by hand with a calculator then to sort through the complex software.

The result is that even when the firm purchases the software, the designers do not use it. This is not a good situation because it means the lighting is laid out with either no calculations or with cursory ones. The result is a lighting scheme that is not only irregular but also wasteful of energy. In addition it means an increased potential for conflict with the client after the completion of the work because of poor lighting design.

This same scenario applies to voltage drop, short circuit, and major equipment sizing programs as well. The software manufacturers have built so many elaborate features into the programs—features that are helpful to sophisticated lighting designers but not to the hit-and-run designer under a deadline—that the programs end up being useless.

The bottom line is that electrical engineers have not yet benefited from the computer revolution. A few tools will be offered here, programs that can be easily and quickly input to build a modest application that will not only create a handy analysis tool but start an education in computer literacy and programming.

The Platform

Assuming a really basic understanding of computer programming, the most sensible software to learn how to use is a database. The language is simple and straightforward for one thing, plus the latest programs do a whole lot of the programming automatically. In addition, there are quite a number of other uses of a good database—in fact, you will find that many of the office and small business applications are database applications.

A quality database program to use is dBase. The latest version is quite sophisticated and has all the tools needed to create some extremely elaborate programs. Crystal reports are built in, along with visual interactive programming, windows programming, and lots of other features that probably won't mean much when you are just getting started—but you will appreciate them when you need them. The routines presented here are from dBase IV, version 2.0. This is a DOS package. There is a new Windows version on the market, but the basic approach and operations are the same. The DOS version will work in either environment.

The first task is to create some databases that will hold all of the manufacturer's data. This is the information that is in the tables normally used to look up all of the characteristics of the conductors. For

TABLE 13.1 Building a Typical dBase File*

Num	Field name†	Field type	Width	Dec	Index
1	SIZE	Character	5		Y
2	PWRFCT_100	Numeric	4	0	N
3	PWRFCT_090	Numeric	4	0	N
4	PWRFCT_080	Numeric	4	0	N
5	PWRFCT_070	Numeric	4	0	N
6	PWRFCT_060	Numeric	4	0	N

*Database ‖ C:\...voltdrop\CONDUCT1 ‖ Field 1/6   ‖   ‖ Num Enter the field name.
Insert/Delete field:Ctrl-N/Ctrl-U
†Field names begin with a letter and may contain letters, digits, and underscores.
Bytes remaining: 3975

example, the temperature correction factor on the conductors is a function of conductor size and power factor. Notice that the "size" field in Table 13.1 is indexed. This means the database is stored so that a quick search can be made for the record according to a given size. This is how the power factor values are extracted from the tables. A close inspection of the program listing reveals many such data files, thus the wildcard used in the above filename. One of these is selected by the "file" variable, which is the number used in the above filename. Then numbers are extracted from this file for the voltage drop calculations according to the conductor size input.

Similar data is also input for copper and aluminum conductors, which order the information by size and ambient temperature. These files are organized in Table 13.2.

The second task is to build a database to hold all of the values that will be input and calculated. This is where all of the information needed for each calculation is loaded from the product-specific databases. The calculation parameters input by the user will also be stored there, along with the results of the calculations. The database will look something like the structure listed in Table 13.3. The first column is the name of the field, or variable. The names are composed to closely

TABLE 13.2 Conductor Data*

Num	Field name†	Field type	Width	Dec	Index
1	SIZE	Character	5		Y
2	TEMP_60	Numeric	3	0	N
3	TEMP_75	Numeric	3	0	N
4	TEMP_90	Numeric	3	0	N

*Database ‖ C:\barx\voltdrop\COPPER ‖ Field 1/4   ‖   ‖ Num Enter the field name.
Insert/Delete field:Ctrl-N/Ctrl-U
†Field names begin with a letter and may contain letters, digits, and underscores.
Bytes remaining: 3986

TABLE 13.3 The Main Program dBase Structure*

Num	Field name†	Field type	Width	Dec	Index
1	FEEDER_ID	Character	20		Y
2	FEEDER_PF	Character	3		N
3	LENGTH	Numeric	4	0	N
4	SIZE	Character	5		N
5	LOAD	Numeric	5	0	N
6	QUANTITY	Numeric	1	0	N
7	VOLT_DROP	Numeric	6	2	N
8	PERCENT	Numeric	5	2	N
9	VOLTAGE	Numeric	3	0	N
10	TESTIT	Character	1		N
11	CONDUIT	Character	12		N
12	METAL	Character	8		N
13	PHASE	Character	6		N
14	TEMP	Character	3		N
15	AMPS	Numeric	8	0	N
16	ROOM_TEMP	Character	7		N

*Database ‖ C:\...voltdrop\VOLTDROP ‖ Field 1/18 ‖ ‖ Num Enter the field name. Insert/Delete field:Ctrl-N/Ctrl-U
†Field names begin with a letter and may contain letters, digits, and underscores.
Bytes remaining: 3891

match the actual value represented (e.g., Conductors for the number of conductors in the conduit, Amps for the allowable loading, and so forth). It is important to assign the field names so they can be readily recognizable while composing, reading, and debugging the code itself.

Notice the naming convention is used not only for all field names but also for the files themselves. In the case of the conduit files the name is set up so that it can be selected on the basis of the last digit, which makes the calculations easier to organize within the program itself. It is important to spend almost as much time organizing the data and naming the variables and files as in creating the program itself. The more carefully the database is organized, the easier the data can be accessed (i.e., with less programming commands, and simpler routines) and the faster and less error prone the routine.

The Data Entry Form

Once the data files are created and the tables loaded into them, it is time to create the screen on the computer from which the calculations will be performed. Again, it is important to organize all of the information in a simple, direct fashion. A good rule is to have the user input information as it is logically collected or calculated. Abbreviated instructions are also needed on the form to identify the keys needed to calculate values, move through the database, and do other important functions.

The general philosophy behind entry form design is to (1) create an efficient, uncluttered screen, (2) provide all of the information that the designer needs to make the sizing or other important decision, (3) provide a simple, direct way for the user to vary all of the important parameters under his or her control and to calculate and recalculate the results based upon these inputs, until a satisfactory result is obtained, and (4) to have the keys designated to perform these iterations readily identified. The latter is important because the user may only use the program every month or two, and it is not productive to have to relearn the program each time it is used. In fact, if that is the case, there is less of an inclination to use it at all.

The dBase package has a form designer, which allows you to create the background and text and then to locate the fields on the form as necessary. Figure 13.1 shows this design screen, which identifies the text that will show up on the form as well as the field names that accept and output data. The field names are the same as those in Table 13.3. Also note that the navigation keys are clearly identified on the form. Additional controls can be built into this form to help the user:

- Help messages identify in a short sentence at the bottom of the page what is required of the user as each field is entered.

- Allowable ranges on data keep the user from entering values that are impossible, unsafe, or otherwise not recommended. If such a value is entered, the program will stop and issue a warning message with the text or instructions provided by the programmer about what is wrong.

```
** Electrical Engineering ─────────────────────────────────────────
 |                        ────────── Voltage Drop Calculations Worksheet ──
 S
 Y   General          CONDUIT TYPE  .. [Steel       ]   ALT-1      Record    1
 S   Information      WIRE TYPE  ..... [Aluminum]       ALT-2
 T                    PHASE TYPE  .... [Three  ]        ALT-3
 E                    TEMPERATURE ... [75 ]             ALT-4
 M
 S
     Feeder           IDENTIFICATION ... [SB/PHDEB              ]
 A   Information      LENGTH (FT) ...... [ 350]
 N                    # OF WIRES ....... [2]
 A                    VOLTAGE   ......... [480]
 L                    LOAD (AMPS) ...... [ 800]  [  760]  ..... AMPS RATING
 Y                    POWER FACTOR ..... [100]
 S                    SIZE (AWG) ....... [500  ]
 I
 S   Special          # OF CONDUCTORS .. [less than 3]
     Conditions       LOADED TO RATING . [N]
                      ROOM TEMPERATURE . [normal ]
                                         [?]  <─┘ to accept values to file
 └──────── Calculated Voltage Drop ........   0.54 V ... 2.25 %  (press END) ──┘
          Describe the section as first/second identifying point.
** Electrical Engineering ─────────────────────────────────────────
 |                        ────────── Voltage Drop Calculations Worksheet ──┐
```

Figure 13.1 Main program calculation screen.

- Limit input fields to specific values such as wire sizes.
- Toggle between fixed values, as in the upper fields on the form.

All of these features make the whole operation very helpful and friendly. The on-line instructions help the user relearn the program with each use, and the data input ranges help to guide the engineer into making reasonable, safe design decisions.

The most important feature of the format screen is that the results of the calculated values are written on the screen after they are calculated. Pressing the <END> key takes the values from the screen and executes the "calculate" subroutine in the program listing. This computes the voltage drop and displays it at the bottom of the screen. After reviewing the results, the designer can cursor through the form and change a value or two and recalculate until the sought-after voltage drop is obtained.

Paging down to a fresh form will clear most of the fields. Several fields are not cleared so that the design can be consistent—for example, in wire type, room temperature, or other values. These values can, of course, be changed from the given values. Also the user can page through the entire database, looking at all the previous records as framed in this same form. This is done by using the <PgDn> and <PgUp> keys.

All of the operations described thus far are either built into the dBase program as automatic features, or they are very simply programmed from within the form design screen. Other features are colors, line drawing, and text font editing. These too are very important in creating a form that is easy to understand and appealing to the senses. Once the form is created, upon exiting, the whole structure is automatically programmed and compiled by the dBase software. The form can be modified, edited, and recompiled as often as necessary.

The Program

The code for setting up the calculation screen and performing the calculations is given in Fig. 13.2. The listing has many embedded comments (lines that begin with an asterisk are nonexecutable remark lines) to explain the various command segments. Even for advanced programmers this is a good idea. First, it helps in writing the code itself. Often a text outline is written first explaining the tasks to be done at each point in the program. Then the actual commands are inserted. When the program is edited—sometimes many years later, as in the case of this listing—it is easy to follow the logical flow of the original routine.

The program itself is divided into several distinct sections:

```
*****
* VOLTAGES.PRG routine to perform voltage drop calculations
*****

PARAMETERS job

*Created 1-23-92
*by William H. Clark II, P.E.
*Bar-X Software Development Co.
*Modified 6-15-94

*Analysis modeled on BUSMAN catalog procedure and data tables

*Public variables
PUBLIC s,file,nofile,xconduit,xmetal,xphase,xtemp,conduits,loads,temps,;
       Cfactor,Lfactor,Tfactor,okamps,q,load
file = 1

*SET system defaults and parameters.
SET NEAR on
SET CLOCK off
SET STATUS off
SET SCOREBOARD off
SET INTENSITY off
SET BELL off

DO CASE
   CASE job = 1
        USE &Bfile EXCLUSIVE
        SET FORMAT TO vtg_drop
        EDIT nomenu
   CASE job = 3
        USE &Bfile EXCLUSIVE
        BROWSE
   CASE job =  2

*KEY commands
ON KEY LABEL end DO calculate
ON KEY LABEL alt-1 DO conduits
ON KEY LABEL alt-2 DO metals
ON KEY LABEL alt-3 DO phases
ON KEY LABEL alt-4 DO temperatures

*Program listing
*ON ERROR
SELECT 1
USE conduct1 ORDER size   EXCLUSIVE
SELECT 2
USE conduct2 ORDER size EXCLUSIVE
SELECT 3
USE conduct3 ORDER size EXCLUSIVE
SELECT 4
USE conduct4 ORDER size EXCLUSIVE
SELECT 5
USE conduct5 ORDER size EXCLUSIVE
SELECT 6
USE conduct6 ORDER size EXCLUSIVE
```

Figure 13.2 Typical computer code listing.

- Accepting input from the operator
- Establishing the environmental settings, variables, and files
- Assigning actions to be performed when specific keys are pressed
- The execution of the main program itself, which targets a specific database in a specific on screen format

Otherwise the code is concerned mostly with extracting the necessary

```
SELECT 10
USE aluminum ORDER size EXCLUSIVE
SELECT 11
USE copper ORDER size EXCLUSIVE
SELECT 12
USE defaults EXCLUSIVE

xphase    = DEFAULTS->xphase
xconduit  = DEFAULTS->xconduit
xtemp     = DEFAULTS->xtemp
xmetal    = DEFAULTS->xmetal

SELECT 13
USE vfactors ORDER size
SELECT 9
GO top
SET FORMAT TO vtg_drop
EDIT nomenu NOAPPEND
CLOSE DATABASES
PRIVATE ALL
RELEASE ALL
ON KEY LABEL end
ON KEY LABEL alt-1
ON KEY LABEL alt-2
ON KEY LABEL alt-3
ON KEY LABEL alt-4

ENDCASE
RETURN
***
PROCEDURE calculate
***

*First data is read into local variables for use later in the routine. Then
*the CASE statement takes the correct value for the voltage loss correction
*factor from CONDUCTR.DBF. This number is a function of wire size and the
*lagging power factor.

*-move cursor to be sure current values on screen are active

*xrow = ROW()
*DO CASE
*     CASE xrow = 9
*       *KEYBOARD (CTRL X)
*ENDCASE

pf = VOLTDROP->feeder_pf
l  = VOLTDROP->length
a  = VOLTDROP->load
q  = VOLTDROP->quantity
s  = VOLTDROP->size
v  = VOLTDROP->voltage
conduits = VOLTDROP->conductors
loads = VOLTDROP->loading
temps = VOLTDROP->room_temp
nofile = STR(file,2,0)
SELECT &nofile
```

Figure 13.2 (*Continued*)

values from the permanent databases on file (i.e., the tables from the equipment catalog) and using these in the performance of the calculations.

This particular voltage drop routine was selected because it describes a very common calculation procedure done by electrical designers. The code commands are simple enough so that it does not take too much imagination to understand, for example, that SELECT 8 means to choose a database in AREA 8. This is exactly the same as

```
            N3 = pwrfct_090
        CASE pf = "80 "
            N3 = pwrfct_080
        CASE pf = "70 "
            N3 = pwrfct_070
        CASE pf = "60 "
            N3 = pwrfct_060
ENDCASE

*-Number of conductors in raceway or conduit factor

Cfactor = 1
DO CASE
    CASE conduits = "less than 3"
        Cfactor = 1.0
    CASE conduits = "4 to 6"
        Cfactor = 0.8
    CASE conduits = "7 to 24"
        Cfactor = 0.7
    CASE conduits = "25 to 42"
        Cfactor = 0.6
    CASE conduits = "43 and over"
        Cfactor = 0.5
ENDCASE

*-Room temperature affect on ampacity factor

Lfactor = 1
DO CASE
    CASE xtemp = 1
        DO CASE
            CASE temps = "88-104"
                Lfactor = 0.82
            CASE temps = "105-113"
                Lfactor = 0.71
            CASE temps = "114-122"
                Lfactor = 0.58
        ENDCASE
    CASE xtemp = 2
        DO CASE
            CASE temps = "88-104"
                Lfactor = 0.88
            CASE temps = "105-113"
                Lfactor = 0.82
            CASE temps = "114-122"
                Lfactor = 0.75
            CASE temps = "123-141"
                Lfactor = 0.58
            CASE temps = "142-158"
                Lfactor = 0.35
        ENDCASE
    CASE xtemp = 3
        DO CASE
            CASE temps = "88-104"
                Lfactor = 0.91
            CASE temps = "105-113"
                Lfactor = 0.87
```

Figure 13.2 (*Continued*)

turning to the data table on page 8 of the design manual, as is required when doing the calculation by hand. Likewise, the USE-ORDER command simply says what column of the table to look at when you get to page 8, and the SEEK command means to go to the value in a specific row of that table. This isolates a single number in the table, just like you do when doing it without a computer. This same sequence of operations is done in whatever table is selected, so

```
                CASE temps = "159-176"
                    Lfactor = 0.41
                ENDCASE
ENDCASE
IF VOLTDROP->loading = "N"
    Lfactor = 1.0
ENDIF

*-Calculate temperature voltage loss factor

Tfactor = 1.0
DO CASE
    CASE xtemp = 1
        Tfactor = 1.0
    OTHERWISE
        SELECT 13
        SEEK s
        DO CASE
            CASE pf = "100"
                Tfactor = pwrfct_100
            CASE pf = " 90"
                Tfactor = pwrfct_090
            CASE pf = " 80"
                Tfactor = pwrfct_080
            CASE pf = " 70"
                Tfactor = pwrfct_070
            CASE pf = " 60"
                Tfactor = pwrfct_060
        ENDCASE

        SELECT 9
ENDCASE

*-If 90F wire voltage loss is twice that of 75F wire

IF xtemp = 3
        Tfactor = Tfactor*2
ENDIF

N1 = l*a
N2 = N1*N3
N4 = N2/(1000000*q)
N5 = (N4/v)*100
N4 = N4*Tfactor
REPLACE VOLTDROP->volt_drop WITH N4
REPLACE VOLTDROP->percent WITH N5
@ 23,42 SAY N4 PICTURE "999.99"
@ 23,54 SAY N5 PICTURE "99.9"

DO CASE
    CASE xmetal = 1
        SELECT 11
    CASE xmetal = 2
        SELECT 10
ENDCASE
SEEK s
DO CASE
```

Figure 13.2 (*Continued*)

the CASE and ENDCASE commands target the correct table. Then all that is required is to simply repeat the above sequence of actions.

The nice thing about the computer is that this series of operations can be performed effortlessly, time and again. The computer does all of the iterations, searches, and page flipping, relieving the operator of this frustrating, error-prone task and letting the user concentrate on the decision at hand: what size feeder to use. Otherwise, the task is so tedious that the designer typically will do only one iteration, and if it

```
ENDCASE
okamps = okamps*q*Cfactor*Lfactor
SELECT 9
REPLACE VOLTDROP->amps WITH okamps
@ 13,51 SAY okamps PICTURE "99999"

DO CASE
    CASE xmetal = 1
        REPLACE VOLTDROP->metal WITH "Aluminum"
        @ 4,35 SAY "Copper"
    CASE xmetal = 2
        REPLACE VOLTDROP->metal WITH "Copper"
        @ 4,35 SAY "Aluminum"
ENDCASE
DO CASE
    CASE xconduit = 1
        REPLACE VOLTDROP->conduit WITH "Steel"
        @ 3,35 SAY "Steel"
    CASE xconduit = 2
        REPLACE VOLTDROP->conduit WITH "Non-Magnetic"
        @ 3,35 SAY "Non-Magnetic"
ENDCASE
DO CASE
    CASE xphase = 1
        REPLACE VOLTDROP->phase with "Single"
        @ 5,35 SAY "Single"
    CASE xphase = 3
        REPLACE VOLTDROP->phase WITH "Three"
        @ 5,35 SAY "Three "
ENDCASE
DO CASE
    CASE xtemp = 1
        REPLACE VOLTDROP->temp WITH "60"
        @ 6,35 SAY "60"
    CASE xtemp = 2
        REPLACE VOLTDROP->temp WITH "75"
        @ 6,35 SAY "75"
    CASE xtemp = 3
        REPLACE VOLTDROP->temp WITH "90"
        @ 6,35 SAY "90"
ENDCASE
SELECT 9
RETURN

PROCEDURE conduits

*-changes conduit material setting

DO CASE
    CASE xconduit = 1
        file = file + 1
        xconduit = 2
        @ 3,35 SAY "Non-Magnetic"
        REPLACE VOLTDROP->conduit WITH "Non-Magnetic"
    CASE xconduit = 2
        file = file - 1
        xconduit = 1
```

Figure 13.2 (*Continued*)

is close, go on further. With the aid of a computer, several trials can be done, in less time, and a more accurate and less costly design can be created, with a greater confidence and safety factor as well.

This is the reason why an effort was made to make the actual screen on the computer as simple as possible and to automate some of the operations with the ALT keys. The easier it is for the user to make repeated trials, the more design iterations that will actually be made, and the better the results will be.

```
RETURN

PROCEDURE metals

*-changes wire material setting

DO CASE
    CASE xmetal = 1
        file = file + 2
        xmetal = 2
        @ 4,35 SAY "Aluminum"
        REPLACE VOLTDROP->metal WITH "Aluminum"
    CASE xmetal = 2
        file = file - 2
        xmetal = 1
        @ 4,35 SAY "Copper  "
        REPLACE VOLTDROP->metal WITH "Copper"
ENDCASE
SELECT 12
REPLACE DEFAULTS->xmetal WITH M->xmetal
SELECT 9
RETURN

PROCEDURE phases

*-changes phases of conductor setting

DO CASE
    CASE xphase = 1
        file = file - 4
        xphase = 3
        @ 5,35 SAY "Three "
        REPLACE VOLTDROP->phase WITH "Three"
    CASE xphase = 3
        file = file + 4
        xphase = 1
        @ 5,35 SAY "Single"
        REPLACE VOLTDROP->phase WITH "Single"
ENDCASE
SELECT 12
REPLACE DEFAULTS->xphase WITH M->xphase
SELECT 9
RETURN

PROCEDURE temperatures

DO CASE
    CASE xtemp = 1
        xtemp = 2
        @ 6,35 SAY "75"
        REPLACE VOLTDROP->temp WITH "75"
    CASE xtemp = 2
        xtemp = 3
        @ 6,35 SAY "90"
        REPLACE VOLTDROP->temp WITH "90"
    CASE xtemp = 3
        xtemp = 1
```

Figure 13.2 (*Continued*)

Applications

For one free computer program send proof of purchase of this book plus $10.95 for shipping and handling ($20.95 if you are overseas) to

Bar X Software Development Company

21203 Paseo de Vaca, Lago Vista TX 78645

or you can order one of the programs free from our Web site at

http://www.netcom.com/nwhcii/index.htm

It will be delivered by email.

The software is a registered, licensed product of Bar X Software and is fully warranted. The book publishers assume no liability or responsibility for the products.

Three programs are available:

- Zonal cavity lighting calculations
- Voltage drop evaluation
- Short circuit analysis

All programs require a 486 or better machine, at least 16 Mbytes of RAM, and 4 Mbytes on the hard disk. They have virtual memory managers for DOS applications. They run in Windows too. They are stand-alone 32-bit applications with a full range of help screens, data entry checks, and printout capabilities.

Chapter 14

Value Engineering

The concept of value engineering has a negative connotation in most contracting circles. It is perceived as the slashing of budgets, quality, and sometimes even safety. Engineers see it as a practice that escalates their liability, reduces their "comfort factor," and causes a great deal of costly redesign time. Owners see it quite differently. Value engineering to them is a tool to guarantee the highest-quality facility for the money. It is their best way to ensure that nobody exacts excessive profits from the project. Which perspective is the correct one? Actually, both perspectives are incorrect. In fact, none of the ideas—which are focused on monetary concerns—have anything to do with value engineering at all.

Perhaps the reason why most people consider the fiscal aspects of value engineering to be paramount is that owners only call upon a value engineering (VE) team after the project is way over budget. Although the true objective of the VE team is not to save money, that is how they are used (or abused) by most owners. Then, when the contractors and engineers are found to be at fault on some costly aspects of a project, the owner offers the VE team as the ones to blame. In reality, it is just as likely that the owner is to blame, having made unrealistic demands of the project and having asked for unnecessary frills.

Whatever the circumstances, it is up to the VE team to single out those items that are either unnecessary, overdesigned, or underdesigned, as the case may be. Politics aside, their purpose is to ensure that the project uses appropriate technology at reasonable market values and profit margins. The auditors therefore have several important objectives before looking at the plans and specifications: to *become well versed in the purpose of the building, the needs of the occupants, the expected lifetime, and the owner's budget for operating expenses.* This is accomplished through meetings with the owner plus an exhaustive review of all documents, correspondence, and studies completed by the design team to date. The next step is to look at the drawings in an effort to *thoroughly understand any special requirements mandated by the owner and their consequences in the actual design to date.*

The special needs cause most of the budget problems. The VE team must review them and present any possible alternatives to the owner. Other objectives accomplished in the process are to

- Point out any instances of overdesign, where a simpler and less costly design will suffice
- Isolate engineering systems and architectural components to ascertain if there are modifications that will make the design work better and at less expense
- Evaluate the interaction between systems and seek potential problems, conflicts, and money-saving alternatives.

The last concept is the most important accomplishment of a good VE team. Even in a well-coordinated firm, most of the design work by each discipline is independent of all others. The design team's purpose is to get the design on paper, with no opportunity to look at the whole picture. The conflicts and details have overwhelmed the fundamentals of the design process itself with the micromanagement of engineering and architectural systems.

Value engineering, on the other hand, analyzes each system in the building as a part of the whole. Although the team includes each discipline, the group has enough experience to be able to collectively— and individually—look at the total picture. Consequently, they are able to perceive options for each discipline that might not be apparent to the design team. Knowing the importance of some design aspects to the project objectives, the VE team members can negotiate between the disciplines to allocate cost-saving changes in favor of one discipline, with minor compromises from others.

The result is a long list of possible design changes, with their associated costs and attributes. Several characteristics are paramount:

1. Constructability
2. Aesthetics
3. Life cycle cost
4. Modification to the design program

Different projects will have different weights for each of these factors. However, the final project must still be a building that will allow the owner to accomplish the needs of the facility.

The Process

A very strong recommendation is to involve a VE team early in the evolution of a project. This is especially important if a firm has frequent cost overruns, either because of inefficiencies inherent in their production staff or because of tight market conditions that favor the contractor. A little extra money spent at the schematic or design development phase can avoid several important repercussions of a VE audit:

- Loss of an important client's confidence
- Reduced profitability because of redesign time
- Increased change orders and liability on the project because so many changes were made in a rush at the final hour
- Damaging the team spirit within your own firm from having to suffer under outside oversight

There is nothing quite as humiliating as going through the agony and doubt of a full VE audit.

Just a little self-discipline during the entire design phase of the project reduces potential loss of face or redesign time that a VE audit might incur. Proper documentation minimizes any liability or fee issues, too. You never know which project is to be value engineered—although the likelihood increases as the project gets larger—so it is wise to establish a rigorous approach on all projects.

The issues addressed by value engineering are not going away either; they will only become more pronounced as issues of energy conservation and indoor air quality dominate the building trades. Electrical designers will need to minimize costs along with the other disciplines. Because electrical costs are more than any other discipline on a typical project, owners want them to be reduced accordingly, whether or not they are code or safety driven.

A good way to see the benefits of a departmental VE program, similar to quality assurance, is to review the actual process followed by a formal VE team. QA was at one time a new and largely superfluous discipline. It is now quite important; so too with VE.

The State versus the Architectural/Engineering (A/E) Firm

Consider the design of a high-profile state office building. The plans are nearly complete and the cost estimate is $7 million over the $25 million budgeted. The owner is justifiably upset and hires a VE team to review the plans and specifications. Knowing that previous VE teams on other projects were sympathetic to the firm of record, the owner arranges for a large out-of-town engineering firm to do the VE audit.

Even then, it is difficult to put together a group of respected experts in any region or community without the presence of some interrelationships to make the deliberations contentious. One remedy is to delegate individuals in pairs. One of the pair is the experienced authority, and the other is a junior engineer with new ideas and a drive to establish a reputation. The former will probably be kind to his associates in the firm under review. The latter will be less hesitant and will pepper the kettle with many new and innovative ideas. The key to making this mix work is to establish in the ground rules that no idea is a bad idea, and thus every single idea that is tabled will receive due consideration.

The first step in the process—a few days before the team meets to start the brainstorming—is to distribute plans to each team member so they can become familiar with the project. Everyone, in the privacy of his or her own space and with reference books and other design aids at hand, will be able to develop some preliminary ideas.

The next step is to have a formal meeting with all the parties present: the VE team, the owner, the A/E firm, and all of their subconsultants. The A/E firm gives a complete briefing of the entire project, from inception to the present status. This is likely to be a stultified presentation, stiff with much resentment and foreboding. It is the job of the VE team leader to stir things up a little.

A successful method is to have a group exercise to agree collectively what the basic purpose of the entire project is. A good way to do this is to participate in a simple exercise to quantify the project. This is the development of a maximum of four phrases of two words each, which between them encompass the entire scope of work. This will be a very frustrating experience for all in attendance as they strive to sort out exactly what the purpose of the building is, what the design objectives are, and what importance to give all the ancillary issues. This process can take several hours, but it is time well spent because it accomplishes several important things:

- It is everyone in the room against the VE team leader. Factions disappear. Animosities and anxieties forgotten, everyone does what is most enjoyable: trying to solve a perplexing problem.

- The group struggles to grasp the most fundamental aspects of the whole process. Why construct the building in the first place? What are the owner's expectations? Why design this way instead of that way? The VE team member needs to know all of this reasoning so they can recommend things that will not clash with the owner's basic philosophy.

- The VE team members are probably unaccustomed to the whole process, and they need discipline. They must focus on some very fundamental issues: what is required, what is expected, and what exists. They must not design or analyze but only consider the feasibility and cost of alternatives. Focusing intensely on the fundamentals helps create the proper state of mind.

This is the overall theme of the meeting between owner, VE team, and A/E consultants. It is a challenging task. It helps to structure this discussion with some other exercises more familiar to the attendees. By first itemizing all of the aspects of the whole project, the most fundamental purpose becomes evident by a process of elimination.

Costs versus Functions

An important part of this preliminary information phase is to begin to associate the costs of the project with specific aspects of the building or the facility program. Given the truism that 20 percent of a project's elements are 80 percent of the total cost, a list of the most significant items totaling 80 percent of the cost is generated:

hvac ductwork	Transmits air
Piping	Transports water
Air handling units	Conditions air
Controls	Modulates environment
Chillers	Cools water
Lighting fixtures	Distributes light
Conduit and wire	Moves electricity

These are 20 to 50 percent of the items, in most cases. This list is not limited to the top percentile of cost items. Additional items from the lower half of the cross section can be added if the cost estimate shows them to have high potential costs. Some such items might be

Foundation	Supports structure
Landscaping	Beautifies site
Light shelves	Enhances daylighting

The overall purpose of the meeting is to reduce the entire scope of work to a single idea. The attendees must come up with a function that defines each item, consisting of a single phrase with one verb and one noun. This exercise focuses attention on the fundamentals. Everyone in the meeting can do this easily for aspects of his or her own discipline. Doing so, for something familiar, makes it easier to develop a two-word function for the entire project.

One final activity is to rank all the functions by cost. If reducing project costs is the objective of the VE audit, it is important to emphasize the most costly items in this group setting. This does not target items for cost cutting but for identifying the significance of the cost. Excessive costs prompt questions from the VE team. For example, a high lighting cost comes from the owner's need to dress up the building with sophisticated luminaries. A high power distribution cost is to reduce harmonics from high anticipated computer density in the building.

The group then rates each item as basic or secondary. Secondary items are not fundamental to the purpose of the building. The building program may modify these definitions. A backup chiller and emergency generator may be superfluous on most buildings. The owner may, however, define it as a basic purpose because of important computer equipment vital to the functioning of all departments.

This process may seem tiresome to those in the meeting, but it is important for two reasons. First, it identifies costly functions that may be basic operations, in which case they cannot be the subjects of design modifications for cost reduction. Second, the process advances the purpose of the seminar: identifying the most basic functions, in pursuit of the most basic function of all.

A logical closure to this phase of the discussion is the codes contingent upon each discipline. The most basic function of the architect is to comply with the building code. The electrical engineer complies with the NEC. In addition, the owner may have adopted a specific energy conservation standard. These technical requirements are paramount. Anything beyond is secondary, unless so identified by the owner.

Diagram of the Plan

So far, the group has distilled the building elements down to a few basic functions. A graphical way to represent this information is to tabulate the times from right to left by the following rules:

1. >>How: The item to the right explains how the item to the left is done.
2. <<Why: The item to the left explains how the item to the right is done.
3. ^^When: The temporal sequence of operations.

As an example, consider the diagram of two-word functions

House government   Satisfy program

These answer the questions

1. How is government housed? By satisfying the program requirements of the agency.
2. Why satisfy the program? So that government will be housed successfully.

The intent of the entire day of discussions is to arrive at the basic purpose of the entire project. The focus of each person has shifted from a particular discipline and the need to fulfill all of the code requirements—and meet the owner's extra program needs. The only question that remains is Why has the A/E design team been assembled? The answer is To build a facility to satisfy the program needs of the owner.

All else follows from this conclusion. The focus is no longer the architect's desire to build a beautiful landmark, electrical engineer's accountability for the fancy lighting or the elaborate raceway system, or the mechanical engineer's need to justify the hvac system and the hot water system. These are now ancillary functions, secondary to the program requirements. If the code does not mandate a particular design aspect, its purpose must be the evaluated.

Now that the basic purpose for the building is established, it is time to review. All written and unwritten program requirements are exposed. Specific design peccadilloes are in evidence. The costs of these questionable items are available, and the team has established their relative worth to the owner and the A/E team. Each VE team member will have a shopping list of questionable items in his or her discipline, plus desirable items to negotiate with the other disciplines. The electrical engineer, for example, may need a larger electrical closet or one closer to the center of the building. The mechanical engineer may want to put AC equipment on the roof. They will both have to negotiate with the architect, but under different circumstances than on the usual project. The value of the changes, in terms of cost and construction, will prevail in this setting.

(This is the benefit of value engineering a job in the early design stages. Often the mechanical and electrical engineers get their equipment rooms where they are available, not where they are needed. Value engineering puts the negotiations on a level playing field, and costs and functions override the benefit of any single discipline.)

The information phase is now completed. The group meeting has exposed all of the conflicting requirements. The VE team leaves the meeting with a better understanding of the owner's needs and the A/E team's dilemmas. They must now exercise careful judgment in the selection of design alternatives to present to the design team and the owner. These ideas (1) must not violate the program, (2) should not be trivial in cost, and (3) should not increase construction time or increase the life cycle cost of the building. The next phase of the process begins the filtering and evaluation of ideas.

Individual Deliberations

The VE team now meets independently for the first time. Every shred of documentation on the project is on hand, and they spend a few hours absorbing as many of the minutiae of the project as possible. The team leader tells them what to look for and continues to guide them away from design criticism and toward conceptual evaluation. Some essential issues to keep focused on at this time include

- Can the function be performed another way?
- Can the function be performed somewhere else?
- How else can the function be done?

Again, the evaluators are not to judge or criticize; they are only to speculate on possible alternatives. Creativity is the only prerequisite to success, as measured by the number of ideas. The only requirements are that the ideas be technically feasible using current, proven technology that can be bid competitively on the open market.

So far, all deliberations have been made by the team members individually, in the privacy of their own minds. They sort ideas by function on a worksheet. This is an important formative process. Everyone is in the same space, supporting each other with project and discipline interpretations. The team members discuss tentative ideas in this informal context and gain positive reinforcement from one another. When others criticize an idea, the team leader is alert to cut criticism short. No idea is a bad idea. The program allows individual prejudices and preferences, but not just yet. Collective reinforcement and support fosters a supportive, nourishing atmosphere so that each concept is able to flourish.

This short, intense phase begins to develop a team spirit. The task itself is so difficult that there is no room for strong preferences or hidden agendas. No one succeeds unless everyone works together. Each idea must not only be good and reasonable, it must also be sold to the owner and the A/E design team. The sponsor of each idea, then, sells the idea to the team. This is preparation for doing so before the real skeptics: the original design team.

Brainstorming

Now it is time for each team member to go public with his or her ideas. The VE team leader has a large easel with a giant pad of paper and a magic marker. The leader writes each idea down and posts the pages on the walls of the room. It is important that the ideas are posted for all to see, rather than just written down on a piece of paper or input to a laptop computer. As the session advances, the members can build new ideas from the ones they see on the board or the walls or verify if an idea is original or not. The growing list becomes the focus of attention. The challenge is to add to it, not to criticize any given idea. The members no longer cut short new ideas, but encourage them. The group process continues until all concepts are exhausted.

The individual dynamics are important, too. It is one thing to have a nice idea on paper, even to chat about it with others around a small group. It is another thing altogether to feel strongly enough about it to present it as a viable alternative to be written on the task board. In so doing, the individual is not only volunteering his or her own time to develop the idea further but is also requiring some time from all members of the group—and later the design team and the owner's time. This is no small matter, considering the limited time available, the high qualifications of all the individuals involved, and the many other demands on their time.

The more time that is wasted on spurious concepts or alternatives, the less time can be devoted to quality ones and the lower the quality of the final recommendations to the owner. No matter how mature the participants, they must collectively prioritize the list so they spend valuable time and resources only on those proposals with the best potential. This is a group responsibility. The team members must put aside their individual priorities, heightened in this speculation process, and dampen individual wills in favor of accomplishing the group objectives.

Evaluation

The team is ready to enter the evaluation phase (see Fig. 14.1). Notice that the tasks described here comprise the workshop portion of the VE audit. The team members participate in only this part. The management and the VE team leader would have been involved in the selection interviews with the client. Then, after the final presentation the team members are needed only for occasional consultation on certain aspects of their recommendations. Also, the owner may request further study on particular projects, for which the services of that team member would be retained for a short while longer. Otherwise, the VE team leader coordinates with the owner to select acceptable projects and then to see to their actual implementation.

In the evaluation phase, the team starts by agreeing about a set of criteria by which to rank all of the suggestions. These criteria come

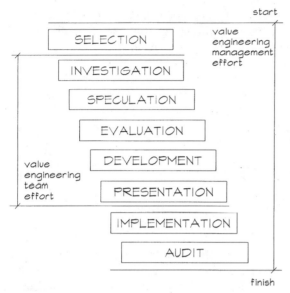

Figure 14.1 Value engineering task sequence.

from the previous discussions with the owner and the A/E design team. For example, the aesthetics of the building may be quite important because it is in a very visible location. In addition, the schedule may be crucial and inflexible. Other important criteria may have evolved through the meetings. Four or five criteria can clearly evaluate the project. The team settles on a reasonable set of touchstones:

1. Constructability
2. Aesthetics
3. Life cycle cost
4. Modification to the design program

The team must now assign a range of possible values to each of these criteria. A reasonable range is 1 through 5. Next, the team must validate the significance of these numbers. The general philosophy is to use 1 as the least-desirable option, 3 as no important differences between the existing system and the proposed, and 5 as an improvement.

The group evaluates all of the proposals by a single criterion at a time. They agree on a rating for the constructability of each item, going through the entire list. It is important for the group to do them all at once so that the ratings are relative and by comparison. It is

TABLE 14.1 Summation of the Project Ratings

Project	Constructability	Aesthetics	Life-cycle cost	Modifications to design program
Two-paned windows	5	5	5	4
2 × 4 hallway lights	5	2	5	4
Water source HPs	3	5	5	1
No light sills	5	2	5	2
Lighting controls	3	4	5	4
More elec. rooms	4	4	5	3
Change bldg. profile	3	1	5	1
Variable air volume	5	3	5	3
No elect. vault	4	4	5	4

also important that the whole group participate in the evaluation process. This prevents any individual from having a disproportionate effect on the deliberations.

The VE team leader jots down the rating numbers next to each item. This is done in print too small for any of the team to read because it might affect their judgment. Then before totaling the four values for each project, the team agrees on a cutoff value. Evaluation is stopped on all projects below this value. This filters the poorest projects from the list in favor of the most productive ones. The result is a group filter on the process of project selection.

A summation of the project ratings is shown in Table 14.1. The highest rated projects are the best prospects according to the criteria established in the first team session. After the members volunteer for the projects (usually the ones they have championed so far), these ratings are a guide to the individual evaluations henceforth. Because each member's time is limited, it must be devoted to the best projects and to those with the greatest potential savings.

As each team member begins to prepare individual project analyses, the VE team leader starts to organize the voluminous amount of information generated so far. A first draft of the audit report will be delivered to the design team and the client at the final group meeting. The purpose of this report is to show that a thorough and complete study was done, as well as to recommend specific projects. While the other members focus on the latter, the team leader focuses on the overall scope of the work.

Table 14.1 is a concise way to impress the client with the team's common grasp of the project objectives as well as all of its subordinate parts. The table lists all the projects conceived by the team, including those below the cutoff number. It is important to include them so that the client is aware that they were considered and then rejected. That way, if implementing all the best projects does not save enough on the project cost, even the second tier of projects may need to be implemented. The table has the overall rating of these projects, too. The owner can begin evaluating them from the highest rating down, until the targeted savings are achieved on the project.

Project Evaluations

The VE team members are charged with preparing a formal report on each of their proposals. An expedient way to organize these reports is to generate a set of forms that have fill-in blanks for all the necessary information. First is a cover page, similar to Fig. 14.2. This gives a title of the project, the estimated savings expected, and the two-word function of the project. This information is penciled in by the team member.

Once the projects are understood in their scope and cost, it is a good idea to contact the appropriate design team member at the A/E firm to discuss the project. There are several possible outcomes of this important conversation:

- The designer had already considered the project and rejected it for compelling reasons. That having been established, the VE auditor will know not to waste any further time or energy on that option.
- The designer had not considered the idea but is supportive of it. This gives the VE team the green light to evaluate it further.
- The designer does not like the idea because it (1) involves new and untested technology or (2) requires a great deal of redesign time. These may or may not be stated, so the auditor—perhaps in consultation with the rest of the team—proceeds in the evaluation process at risk.

Of course, some of the suggestions may be so good that the design team accepts them immediately, without telling the auditing team. That way, when the formal presentation is done, the original designer can say that particular project has already been implemented. If this occurs, instead of being confrontational, the VE team member should take solace in the fact that his or her idea has been implemented and not fret about who gets the credit. It will all come out in the final analysis, when the next cost estimate for the project comes into budget.

Otherwise, it is important to seek feedback from the design team. After all, if a project is presented at the final meeting that really has been implemented, the VE team may suffer in credibility. Also, the client will be left with the impression that the evaluation process is

```
PROJECT NUMBER _____  SAVINGS _____
TITLE _____
TEAM MEMBER _____
FUNCTION _____  _____
              verb                    noun
_____
_____

[ ]  Accepted as presented        [ ]  Rejected for reasons noted
[ ]  Accepted as noted            [ ]  Rejected for reasons to be
                                       provided by _____
                                       on date _____

[ ]  Accepted as noted
[ ]  Acceptance/Rejection deferred until _____
[ ]  Status changed
     From _____  To _____
     By _____  Date _____


_____   _____
Director, Design & Construction   Project Manager


_____
VE Team Coordinator


COMMENTS:
_____
_____
_____
```

Figure 14.2 Project approval form.

something less than a team effort. If this happens, both the design team and the auditing team lose; and they lose not just esteem with the client but perhaps even future work. It is the responsibility of the VE team leader and the project manager to coach their people to ensure such schisms do not happen. If conflicts are inevitable, as is usual when highly experienced professionals coordinate their efforts, the antagonisms should be stage managed by more mature heads to happen in private away from the critical scrutiny of the owner.

Once all of the thorny issues have been resolved—or at least identified—the VE auditor can proceed with the project analyses. All projects will require very thorough explanation and cost evaluation. The potentially controversial ones should be supported even more, time allowing. Sometimes the possible enhancement in value is not worth the effort or the possibility of antagonizing the owner or the design team. In such circumstances it is the individual auditor's responsibility to consult the other team members. There may be other issues involved, completely unrelated to the personal ones perceived.

Notice that more and more the VE group is well advised to act as a team. Without total coordination and trust between them, they risk considerable collective embarrassment. After all, they have had only a few hours to absorb information that the design team had mulled over for many months. It is said that over a million decisions go into the development of a thorough set of construction documents for a sizable project. These decisions will have been carefully made by the project designers. For a few individuals to pick the few decisions among these that affect value disproportionately to cost is a task requiring teamwork at its best.

In the midst of these political and personal ramifications, it is almost comforting for the team members to have a set of carefully organized documents to fill out, not to mention having some structured help in organizing their thoughts. A set of four blank pages after the cover page are for just this purpose. These are as follows:

1. A lined page, titled "Notes and Discussion." This begs a written discussion of the project, its advantages, disadvantages, and general scope.
2. A blank page for a sketch of the design to accompany the written description.
3. A second blank page to illustrate the recommended alternative design.
4. A cost comparison page, listing each major item of the project, with before and after costs (see Fig. 14.3).

The next page (Fig. 14.4) requires the auditor to do a little engineering economics to calculate the life cycle cost for those projects that differ in operational costs. Many projects will involve some redesign costs, as noted in the top column. Otherwise the rest of the daunting form is left blank unless (1) the project changes power consumption considerably, (2) there are marked changes in maintenance, or (3)

Value Engineering

Item	Description of Modifications	Costs		Trade-Offs
		Before	After	
Grand total/page:				

Figure 14.3 Project comparison form.

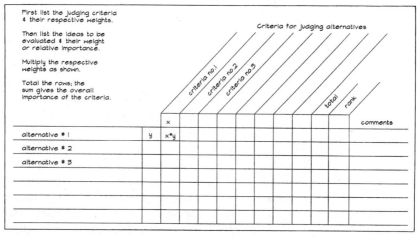

Figure 14.4 Options matrix presentation.

there are significant changes in the frequency of replacement. These eventualities should be ignored if they are not consequential.

The last page of the project report (Fig. 14.5) summarizes the recommendations by the VE team. "Strongly recommends," followed by a brief explanation is all that is needed in most cases. Then the cost is written in the appropriate blank, according to the type of project:

I   The project does not affect the building function of essential characteristics.
II The building essentials must be restated as different from the owner's basic program, but the highest-order functions are unaffected.
III Some reduction in scope is required, such as staging the development of the project (e.g., shelling out one floor or building a wing later).

The VE team leader, when all of the projects are assembled in the workbook, will calculate the percentages on this form.

There are two other important inputs for this form. First is the identification of any related projects or modifications of the proposed project that will enhance the value of the project. Next are any special comments that identify changes in the original construction of the project, such as scheduling concerns or potential conflicts with other disciplines.

Finally, there is a lined page titled "Outline for Oral Presentation." The team member can outline the presentation and the highlights to be made and note any other pertinent information. Copies of all the other pages of the report will be sent to the owner and the design team; this one is for the team member alone.

The Presentation

After many days of long, stressful hours in close quarters, the VE team must now brush off their collective weariness and their individual differences. They must muster the fortitude to present a united front—despite, perhaps, having only made each other's acquaintance a mere week earlier. They need to find the enthusiasm within to present the projects in an adversarial context.

The single most important goal is for the group to present their work as a team. This can be made clear in the VE team leader's introduction to the session. Even though individuals present specific projects, for which they did much of the research and analysis, each project has earned the support and approval of the team. The projects were first evaluated and approved by the group and then assigned to an individual for further evaluation.

A good way to further the group concept, throughout the presentations, is for other team members to offer evidence and support to the current presenter. This will impress the audience with a strong front of group unity, and each project will have more weight as a result.

```
┌─────────────────────────────────────────────────────────────┐
│  Other opportunities for value improvement:                 │
│  _____  │
│  _____  │
│  _____  │
│  _____  │
│  _____  │
├─────────────────────────────────────────────────────────────┤
│  Summary of savings:                                        │
│                                                             │
│        Category I     = _____ or _____ % of total project │
│        Category II    = _____ or _____ % of total project │
│        Category III   = _____ or _____ % of total project │
│                                                             │
│        Total Potential                                      │
│        Savings Identified: = $ _____ or ___ % of the total project │
├─────────────────────────────────────────────────────────────┤
│  Other opportunities for value improvement:                 │
│  _____  │
│  _____  │
├─────────────────────────────────────────────────────────────┤
│  Implementation plan:                                       │
│  _____  │
│  _____  │
│  _____  │
│  _____  │
└─────────────────────────────────────────────────────────────┘
```

Figure 14.5 Project evaluation summary.

This is especially important for the more controversial projects. It is likely that these, by their nature limited to specific disciplines, will have a single detractor on the A/E design team. This individual may be more open to ideas perceived as coming from a united group than from a single opinionated member.

Otherwise, the whole meeting is carefully structured to facilitate the exchange of ideas and to ensure that the most information possible is exchanged because this meeting is, in fact, still a part of the value engineering process. It is a forum for the presenter's ideas as well as a place for the design team and the owner to provide further substantive input on these ideas. The project recommendations presented by the team members are only tentative; the final selections are not determined until after the final presentations to the owner and the design team.

As each person arrives at the meeting, he or she is given a large binder with the draft report for the VE audit. This report has several sections, in the following order:

1. A project statement summarizing the purpose, use, and occupancy of the building evaluated. Any special systems are described too.
2. A list of the team members, their firms, and telephone numbers and with resumes in an appendix at the end of the report.
3. The word-processed summary of all the projects considered by the group, along with the grading scheme and the score for each.
4. The reports on each project, with the five- or six-page evaluation form filled in by hand of all of the pertinent information developed by the auditor.
5. Appendixes detailing any special or unusual standards the design must comply with, the formal cost estimate prepared to date (if such is available), plus graphical presentations of the costs for each phase, such as provided in Tables 14.2 and 14.3. For very large projects, a cost model for each discipline will also be provided.

Lessons for In-House VE Programs

There is one very important lesson that VE has in the production of engineering plans and specifications. It is too late when an owner hires a VE team to conduct an audit. Such oversight is not considered detrimental by many large A/E firms. They perceive it as a necessary part of the process. In fact, they should consider a VE audit in the same light as an audit by the IRS. In oriental terms, having to submit to an engineering audit should constitute a great "loss of face," or esteem, among peers.

Most owners pay homage to the very western concept of independent auditing to assure quality and savings. They do so by paying for the expense rather than writing into their contracts a qualification to charge the design team with the task of value as well as quality control. This is the best way to ensure quality and professionalism.

TABLE 14.2 Total Construction Costs by Category

Description	Cost ($)	%	Cum. %
Finishes	5,578,967	19.42	19.42
Concrete	4,920,871	17.13	36.56
Heating, ventilating & AC	4,225,647	14.71	51.27
Electrical	4,001,439	13.93	65.20
Sitework	1,849,287	6.44	71.64
Doors & windows	1,719,314	5.99	77.63
Masonry	1,622,532	5.65	83.28
Conveying	900,550	3.14	86.41
Metals	899,048	3.13	89.54
Plumbing	742,975	2.59	92.13
Special construction	643,950	2.24	94.37
Thermal & moisture protection	546,397	1.90	96.28
Fire protection	522,549	1.82	98.09
Specialties	187,931	0.65	98.75
Wood & plastics	159,370	0.55	99.30
Furnishings	155,666	0.54	99.85
Equipment	44,032	0.15	100.00
Subtotal	28,720,795		
General conditions @ 5%	1,436,040		
General contractor fee @ 5%	904,705		
Design contingency @ 2%	621,231		
Total	31,682,771		

TABLE 14.3 Electrical Costs

Description	Cost ($)	%	Cum. %	
Conduit and wire	1,065,545	26.58	26.58	
Equipment	1,031,663	25.78	52.36	
Lighting systems	856,596	21.41	73.77	...80% of costs
Systems	727,829	18.19	91.96	
Devices	321,806	8.04	100.00	
Total	4,001,439			

The best engineering project is one that proceeds at a steady pace from start to finish. Engineering and architectural decisions are made deliberately and carefully. Communication among disciplines is a planned and managed process. Team members are told what they need to know when they need to know it.

A VE audit puts this whole program into a state of chaos. Many changes are made in all disciplines at once. Then a crushing deadline is imposed not only to meet the client's target but also to minimize the time spent on redesign so that all of the profit is not wasted. Everyone rushes to finish. Careful calculations are approximated.

Coordination between the disciplines is lacking. The plans are bid in a hurry and many addenda items are issued to clarify inconsistencies.

If VE is to be done at all, it should be an integral part of the whole production process. It should be the design team's responsibility, and they should welcome—even ask for—it. No professional designer is comfortable with the stress and confusion that characterizes the final stages of most large projects. Nor do they enjoy facing an unhappy client or contractor. Even an antagonistic VE team is friendly by comparison. Make them a part of the design process, and soon clients will be happier with the work and the contractors too. The latter will bid your projects with a closer margin, and owners will pay more for the higher-quality work. Calls from the client will be friendly and complimentary instead of questioning and adversarial.

The next rule to follow is to make the VE audit as independent as possible. This limits the interpersonal politics and the petty prejudices and gives the evaluators the freedom to make honest judgments. In a large enough firm, the team can be completely anonymous. Otherwise, using the grading criteria explained above can help to rid the process of any undue influences by individual team members.

Finally, it helps to have a formal, written evaluation. The structure helps the evaluator organize his or her reasoning. It also depersonalizes the process so that the designer being critiqued will not be inclined to consider the comments personal, but more professional and educational. The forms also help management keep informed. Over time, they indicate the scope of the whole review process and can highlight shortcomings in the production department. For example, if the mechanical department always has more problems than the other disciplines, the fault is with the department head or the overall experience of the staff. This connection could not be verified without the paper trail left by the formal evaluation process.

A good overall strategy on the management of the VE audit is to treat it as a training session, with senior members of the firm evaluating the work of junior members. In this regard, the management does the entire firm a valuable service. Their experience is put to best use, and senior staff are also kept informed on the pace of major projects.

Value Engineering Smaller Projects

It is not necessary that VE be limited to large projects. The same principles, in a less formal statement, can be quite helpful on any size project. It is an ideal way to formulate an independent assessment of a project without pointing fingers and ruffling feathers. A formal report is generated as a combination of design review and economic evaluation.

Another significant application of the VE method is on projects already under way, which have reached an impasse. There are many reasons why this might happen; a few of them follow:

- One of the contractors withdraws from the project or declares bankruptcy.
- A major equipment supplier is delayed or provides substandard equipment.
- Existing conditions in a remodel project cause unforeseen problems.
- A change in management in the owners organization shifts priorities.

Any one of these situations can cause havoc on the job site. The problems can then escalate if left unresolved, leading to work delays, cost overruns, and so forth. The project manager and the contractors may make every effort to get back on schedule, effort that would swing any other project right back on course. On this project, though, the workflow is so disrupted that their best efforts fail. Then the owner gets upset enough to get lawyers involved, and then everyone starts losing money and faith simultaneously. At this point, the trust that melds the diverse elements of the design and construction team together is lost, and everyone becomes suspicious of everyone else.

The only way out, short of litigation, is to submit the entire project to an independent evaluation by someone with no personal stake in the outcome. That is, instead of spending precious funds on lawyers, spend it more constructively with the goal of fixing the situation and saving the owner money in the process. Not just saving money on legal fees, but money on construction costs and energy use once the project is completed. This is the strength of the VE process, to refocus the attention of all parties on the technical goals of the work, rather than upsetting everyone by fault finding.

For the owner's edification, the VE process should be presented as a no-lose situation. It holds the possibility of resolving the issues and raising the quality of the completed project. Then, even if this objective fails, the VE audit will have collected valuable and important information for use in any litigation that ensues. That is, it should distinguish between the original design and the contractor's implementation thereof, as the culprit.

An example should help to clarify these subtle points. The situation involves the use of a relatively new technology by a start-up manufacturer and for an owner anxious to conserve energy using the state of the art technology. The technology is a natural gas-driven chiller installed to reduce electric power usage, and the technology is a small 100-ton chiller that is essentially custom made for the job.

The chiller is installed, the system piped in, and powered up. A few problems occur, as is expected for a complete hvac renovation of a 30-year-old building. Valves are installed backward, pump seals rupture, and the pump is unable to develop enough pressure to deliver water to the air handler coils. These are common problems with a major renovation and are usually taken in stride.

The electrical work is done without a hitch. A new variable-speed drive is installed on the air handler and the cooling tower fan, power is routed to the new pump and the new chiller, and controls are wired properly. The control software, which is custom designed by the chiller manufacturer, is found to have a few problems in coordinating the various elements of the chiller—the engine, the condenser, and the glycol pump. Even this is not unusual, and most manufacturers are happy to fix any problems.

Three things happen to disrupt the entire process. First, the chiller manufacturer announces they will be declaring bankruptcy imminently. What little faith the owner might have had in the new technology is dashed. Second, there is a change of management at the owner's business. A strong understanding between the owner and the engineer based on 20 years of friendship is lost, and what little leniency this might have gained the engineer to overcome the project difficulties is lost. Finally, the original engineer of record on the project leaves the consulting firm, and some of the continuity in the design and construction management is lost.

To make matters worse, the system cannot be made to operate with the new chiller in line. The culprit seems to be a chiller that has too high a static pressure, so a new pump with greater capacity is required. A new impeller is installed in the pump, to no avail. A new pump is called for, and everyone squabbles over who is going to pay for it—the engineer, the contractor, or the chiller manufacturer. The owner, who has already involved his lawyers to investigate product warranty issues involving the soon-to-be bankrupt manufacturer, has a change in management. The new overseer asks the lawyers to go one step further and to name the engineering firm in a suit for having specified an unproved chiller manufacturer in the first place (even though the owner had been the one to select the machine and the manufacturer).

All of the difficulties would seem to be mechanical problems on the surface. However, it is the electrical controls and custom-designed software that is operating all of the mechanical equipment. If the controls are not properly configured, the mechanical equipment will not work as designed.

Despite the best efforts of all the parties involved to complete the work in a professional and expeditious manner, the teamwork has dissipated into blaming. All the while, the owner keeps losing money because the mechanical contractor keeps charging for diagnostic tests on systems that are incomplete and perhaps inadequate. Needless to say, it is everyone's worst nightmare (except the lawyers). The entire process is perilously close to going to civil court, to everyone's ultimate loss (except the lawyers). Short of scrapping the entire project, the engineering firm asks for an independent VE audit to take an unbiased view of the design and the installation, on the outside chance of making it all work before the appointed court date.

The VE report is as follows.

```
Project Manager
Clark Engineering

REF: Jax Chiller Replacement Project
```

This is an internal memorandum intended to provide an independent evaluation of the above referenced project. It is more of a value engineering evaluation of the project, instead of a critique. The report is based upon a general knowledge of the work from two brief walk-throughs, a review of the original and the new design drawings, and conversations with the engineering personnel of record, the contractors, and the owner.

It is my perception that the project is too complex for the owner's tolerance. The chiller, as the most problematic item to date, is not necessarily the main problem, just the scapegoat. I believe that if other problems are resolved, the owner's level of satisfaction will increase, and he will get what he perceives is the intrinsic value of the project. Consequently, the recommendations to follow focus upon the need to simplify the project and in the process to eliminate some of the more costly items that are in any event superfluous.

The first complexity is the use of variable frequency drives (VFDs). The air handling unit (AHU) is a constant-volume unit and will see little energy savings with a VFD driving it. The VFD in the cooling tower is more applicable, but I doubt if the energy saved over the use of a much simpler two-speed motor is much because of the constant loads in the building due to very high interior equipment loads and high exhaust requirements (i.e., high outside air loads). Removing these two VFDs will make the system an order of magnitude simpler, easier to control, and less costly to maintain in my opinion.

The second complexity of the controls is the hydronic loop. The existing pump, if possible, should be made to move water through the entire air handling loop (isolating the new chiller) so that the existing balancing valves can be used to modulate the flow to the AHU. This will provide sufficient flow to the operating AHU without resorting to complex, costly, and tentative software programming as a solution. The system operating under these conditions will, furthermore, prove the hydronic side and also prove that the system will perform as designed with the new chiller in the loop.

The third complexity is the air side balance. The full return air path needs to be established via the louvers above the double doors into the print shop from the main building. Then the supply air can be balanced. Until this is done, the static pressure at the AHU will be excessive and the rooftop unit above the small addition will be overloaded because it is operating under a negative pressure and is therefore conditioning more than its designated space load. Once the air side is balanced, the main AHU fan can be sheaved or adjusted to provide the design cubic feet per minute, and the VFD can be removed from the motor.

The fourth complexity is the chilled water loop. The problem seems to be that an inadequate head is available to the loop. The two main sources of static losses are the chiller and the coils. Because the chiller (we are told) has a higher than designed pressure loss, the coils need to make up the difference. The coils, it seems, may not have been cleaned since they were installed (20 years ago) because until this project there was no way to pull the coils (the room was too tight). Dirty coils from 30 years of use will increase air as well as fluid resistance, which both seem to be a problem. The coils are a likely culprit, and not the new chiller, for the unexpectedly high pumping static pressure.

It is important to be sure that all elements of the chilled water loop are in proper working order, including flow through the chiller, before assuming the only solution is a higher-horsepower pump. This may make the system work, but it will not solve the problem. Excess air in the loop could also cause these problems, for which the solution is also oversizing the pump. A bladder-type system should not have this problem, if the closed hydronic loop was properly purged of air upon installation.

The system as designed is a quality installation, designed to optimize energy savings, using the best available technology. Unfortunately, the reality is that a new chiller may be all the complexity the installation can handle. A simplified system will probably use more energy to operate, but it will be easier to diagnose and fix both now and for the useful life of the equipment. (We see just how hard it is to diagnose problems now.)

In the absence of detailed test results, it is not reasonable to make any more comments. The following information should be a priority:

- Pressure drops (air side) across the coil versus manufacturer's data
- Water side pressure losses through the coils, the chiller, and the loop as a whole, all at design flow rates

If the problem is in the AHU coils, they should be pulled and cleaned or repaired as required. The piping arrangement to the coils will need to be reconfigured in any event, so it will be easy to extract and inspect the coils at that time. (If the mechanical contractor has a long-term service contract with the owner, they should be anxious to do this anyway.)

Sincerely,

VE Auditor
Clark Engineering

The key point is that the system is overly complex for the application. The owner may very well have required this level of complexity in the original design meetings—perhaps even to the objections of the engineer—but management has changed and so must the design objectives. A significant portion of the design complexity is in the electrical side of the system: the VFDs, the software control of valves, and the temperature sensors and actuators controlling the flow of air to the spaces. These were not even thought to be a possible culprit until the VE audit was done.

Furthermore, the whole objective of the project was to save on electricity use by using the less-expensive natural gas to drive the chiller compressor. This directive was carried through on all aspects of the project, by the use of variable-speed drives on fans that might have done just as well without them. The original design did not consider the cost of the drives, the controls, and the electronic modulators to control the hvac equipment (the original pneumatic controls were all changed to electronic). Had these electrical implications been evaluated, it would probably not have been possible to justify the higher cost of the equipment in terms of energy savings.

Chapter 15

Load Control and Harmonics

One of the best and most helpful services the electrical designer can provide a building owner is education about the mechanics of the utility billing structure. Most small business owners do not realize the great influence that patterns of energy usage have on the utility bill. Most do not even realize that there are two parts, for straight usage and for peak demand. The electrical engineer who shows the client the difference will be long remembered by the owner.

Motors are the easiest devices to illustrate the parts of the billing cycle. The steady-state operation under relatively constant load is the usage part of the bill. Cycling the motor on and off causes peaks in electricity use, because of the inrush current needed to start the heavy motor from dead stop, under load. It is like a car starting up a steep hill at a traffic signal, in fourth gear. Because the motor has only one speed that is geared to operate in the most efficient overdrive situation, it must struggle mightily to start from a stop. Starting a car from a stop in fourth gear does not work even with the accelerator all the way down. A motor uses up to 4 times its maximum steady-state energy to get started.

This extreme energy use causes a large spike in the energy used at the facility. The effect is even more pronounced if the building power drops off-line, and then starts back up. All the motors in the building start simultaneously—elevators, air handler fans, compressors, chillers, and cooling tower fan motors. All of their inrush currents occur simultaneously. It is no wonder the power is often lost again just after it is first returned, because the current draw is so extreme.

Every time this peak happens, the utility meter records the maximum. That maximum is used to determine the demand portion of the bill, which is usually half or more of the total monthly billing. Some rate structures use the maximum peak for several subsequent months of billing or even a whole year of billing. So, if power is lost only once during business hours, 12 whole months of charges can be affected.

As simple a thing as turning all major equipment off when the power goes out can save thousands of dollars. Conversely, turning major equipment on at 15-min intervals after the outage or first thing in the morning (usually peak demand is measured in 15-min blocks) can save just as much, if not more.

This is a rather crude form of load control. It should convince the owner that an automated system is definitely worth considering. The larger the building, the more savings that are possible through energy management systems.

Electrical Load Profile Elements

The idea of load control has traditionally been in the realm of the hvac designer and is typically an obscure component of the hvac controls contract. This is not always a very effective way to do it. Take as an example two of the most popular and prevalent hvac control techniques, duty cycling and demand limit control.

The principle of duty cycling is to turn off equipment on a specific schedule. A temperature-dependent schedule is based on the concept that as the outside temperature increases in the summer, the electricity consumption by the hvac equipment increases. Turning off major energy users at a certain preset temperature will reduce energy use, according to the theory. Leaving them off for only 15 to 30 min will minimize the effect on the occupants and make the cycling of equipment hardly noticeable.

The problem is that when the ac equipment comes back on line, the building load is much higher than it was. The building temperature is higher, and so the ac equipment must work harder to catch up. Furthermore, because the equipment is going from all stop to full ahead, the load peaks because of inrush current to fan motors, pumps, and condensers. Also, at higher loading conditions (100 percent plus in many such instances), the hvac equipment efficiency drops off drastically. So, overall, actual kilowatthour use may have been reduced, but kilowatt demand use can potentially go up quite a lot.

The second popular hvac load control strategy is demand limiting. This is the process of turning off loads temporarily to keep the load limited to a specific kilowatt level. As the load on the metered point increases toward the preset kilowatt maximum, loads in the building are taken off-line—air handlers, exhaust fans, and elevator motors. This would seem to be the right way to do it, except that when the loads are brought back on-line, the individual systems are typically overloaded. They will have to work longer and at a lower efficiency to compensate for the time lost. Just because the system is not working does not mean the load goes away. When the ac goes off, the space heats up, and the furnishings with it. When the ac comes back on, the space as well as the furnishings must be cooled down. Energy in equals energy out.

The only way duty cycling and demand limiting can save real energy use is if they operate simultaneously, working to limit both kilowatt and kilowatthour usage. This will reduce the efficacy of each and is likely to cause the equipment to cycle frequently. This may in fact save energy, but it will cause extreme wear and tear on the machinery, going constantly to full load to stop and back again. The cost of repairing this equipment more frequently and replacing it twice as often must be factored into the payback analysis of the controls in order to reach a realistic number for the investment.

The bottom line is that a certain amount of energy must be used to maintain the environmental and other vital services in a building. Delaying those services does not save any energy at all and only saves money if the load is delayed so that the power used is billed at a lower, after-hours rate schedule. It may therefore be better to spend the money on high-efficiency motors rather than a fancy controls system. Alternately, programming thermostats to change setpoint as the building reaches peak load will modulate the system so as not to cause undue wear on the equipment and will also reduce peak energy use. It will only reduce total energy use if the temperature remains at the higher value until the close of business.

Proactive Controls

Proactive is a modern term that implies that problems are taken care of at the root cause before they manifest themselves as problems. Reactive has the opposite sense, meaning the problem is already there and is noticeable before any action is really taken to resolve it. The demand limiting and duty cycling strategies are essentially reactive in nature. They may anticipate the problem, but they still allow the problem to develop and mature.

Reactive methods are inappropriate strategies now because the new outside air standards have been implemented for all occupied spaces. Cycling equipment on and off reduces the outside airflow to a space. This is no longer permitted because it diminishes the indoor air quality (IAQ). This applies only to constant-volume systems, which are most common for small and medium-sized commercial applications. It is even worse for variable-volume systems because if the air handler volume is diminished, the relative percentage of outside air must increase because the same net volume of outside air must flow to each space at all times.

The result of these regulations is that the onus of energy conservation is moved from the hvac engineer to the electrical engineer. There are some things the former can still do, but they are far more costly and less effective than the electrical technologies that are available.

The traditional reactive hvac strategies no longer being effective, the only remaining means for many installations is to be proactive. They must strive to reduce the building air conditioning load before it must be dealt with by the hvac equipment. (The same applies to northern climates in the winter. However, the heat source is not electricity, so it is not an issue here.) Several very effective ways the electrical designer can help toward this objective will now be presented.

Lighting

There are proactive as well as reactive ways for proper use of lighting to reduce both the peak demand and use electricity bills. The nice thing about lighting is that there is no substantial inrush current to complicate the issue. The only factor that should be considered in lighting controls is the effect upon rapid cycling of fluorescent lamps, which reduces their useful lifetime. A well-conceived controls scheme and appropriate selection of controller, and its timing, can avoid excessive cycling and its detrimental effects.

The most efficient of all lighting schemes is simply to use daylighting. This is especially appropriate for workstations that have computer terminals in use most of the time. It is not uncommon, for example, to see the blinds closed and the lights turned off in CAD operator areas because the glare on the CRT screens is so distracting. In most other situations even a small amount of natural daylighting, with perhaps the addition of a small low-wattage task light to highlight the work surface, is more than adequate.

The next logical consideration is to provide only enough light in a space as is absolutely necessary: design by the code. Most spaces are 10 to 50 percent overlit, especially spaces with daylighting or task lighting available. A properly designed lighting scheme should never need task lighting, unless it was designed with this purpose in mind. Nor should occupants keep blinds closed on windows all the time, with the lights full on. This scenario may be more the fault of the architect or the interior designer in providing excessive window exposure, but part of the fault must be the lighting designer's, in not advising them of this circumstance.

In any event, the nice thing about reducing lighting is that it simultaneously reduces the hvac load, too (with a small heating penalty in the wintertime). Brightly lit spaces also tend to feel warmer, so there is a little further saving due to subliminal effects.

Another proactive strategy is to use high-efficiency light sources.

This means using fluorescent and other lights, instead of incandescent lamps (which should be used for specialty applications only). It also means using high-efficiency fluorescent lamps and ballasts for all light sources. Finally, it means specifying luminaires with efficient and effective photometrics. They must deliver the most light to the work surface.

There is more to efficient lighting than the selection of efficient and appropriate light sources. Ceiling height and room configuration are important as well. Fluorescent lamps, for example, are far less efficient than metal halide lamps in high ceilings. Spot lights or recessed down lights in high ceilings are superfluous. Pendant-hung fixtures are appropriate in many settings with high ceilings, so long as the up light is minimal. Indirect lighting via ceiling reflectance should be used only for specialty applications such as in libraries, lobbies, or conference rooms where the extra cost and the low efficiency can be justified for aesthetic reasons.

Lighting controls are the final touch. The aforementioned offices with the blinds closed all the time should have multiple-level switching so the occupants can take advantage of available daylighting. High-wattage exterior lights need photocells to keep them from operating when not needed. Large spaces with a high light load and low occupancy—such as store rooms, conference rooms, or dining halls—could have both multiple switching and controls, such as occupancy sensors. This is a judgment call that is sometimes too hard for either the owner or the designer to make. A compromise is to put a lighting controls allowance in the project budget. After the facility is occupied and in a pattern of occupancy, the controls can be installed in appropriate spaces that are often either overlit or unnecessarily lighted.

Motors

Rare is the circumstance in which the most efficient motor should not be specified at a facility. This includes all motors in every application. The only mitigating factors should be very low time of service and service or availability from the manufacturer. Even these are seldom valid, what with the wide selection available from a plethora of reputable manufacturers.

The next rule for motors is to size them exactly to the load or even a little under the load. Any quality motor can operate a little above the curve for short periods of time, without adverse affects. An undersized motor, no matter its application, might as well be the least-efficient model available. Mechanical designers almost always oversize their motors for pumps, fans, and any other application. In most cases the use parameters are already conservative, so the motors are very oversized.

In the design of a building, the electrical designer is in a similar role to a commercial utility company. The electrical service must be designed to handle the peak load for any branch or point of use. The electrical designer, being at the low end of the pecking order, does not have the utility's authority to impose penalties for excessive peak use. The designer must simply design the service to the hvac engineer's specifications.

It is up to the electrical engineer to assert the design priorities on behalf of the client. As such, he or she should have the authority to ensure that each motor and the entire electrical service along with it (so that the peak inrush motor loads can be handled) is not excessively oversized. The best way to do this is to require a load test upon installation. If a motor is operating under full load conditions at less than 85 percent of its rated capacity, it should be changed out to a more appropriately sized motor.

This performance testing will not allow the size of wiring, disconnects, breakers, panels, and service feeders to be reduced. It is already too late for that if the installation is complete. Still it saves money on the motor and on its operating costs. It will also influence the mechanical designer on future projects to be more careful in selections, knowing that they will be tested upon installation.

Equipment

The electrical designer has some control via the specifications over many electrical appliances. These might include the equipment in a kitchen, shop, or restaurant. The electrical designer has control over further equipment by the power of persuasion with the owner. Examples are televisions in hotel guest rooms, computer monitors in an office, refrigerators in efficiency motel units, and almost any other items of major equipment at a facility.

It does not cost much more to purchase the most efficient equipment. In some cases it may cost less because of rebates from the manufacturer or from the local utility company. This is an opportunity to reduce energy costs for the life of the building, and the design engineer would be remiss in not advising a client to take advantage of the very latest technology on the market.

Electric Heating

One of the most easily avoided uses of electricity is in direct space heating. At a minimum, a heat pump is more appropriate than an air conditioner with electric strip heat. If there is no other way around it, the strip heat should be on a separate breaker that can be turned off until absolutely necessary. Even if natural gas is not available, a small tank will fuel a furnace far more cheaply than will electricity. Fuel oil is another option.

A third possibility for medium-sized and larger facilities is to install a hybrid water source heat pump system. A cooling tower is sized to handle the cooling load, and a ground loop is sized for the heating load. In northern climates, a boiler is used for heating and the ground loop for cooling. The more expensive ground loop is used for the smaller of the seasonal loads. This system is particularly cost effective in warm climates that otherwise use straight electric heating.

During the off season, the ground loop is available for (real) peak load reduction. In a warm climate, for example, the ground loop can be used to reduce the cooling load in the summer. In fact, the more heat that is rejected to the ground in the summer, the more heat can be extracted from the ground during the heating season.

Another very effective technology is one that incorporates hot water heating with the air conditioning heat pump. In the summer, heat from a building is transferred directly to the hot water, instead of to the air via an outside condensing unit. These devices are no longer limited to low-demand residential applications but have sturdy commercial designs as well. These industrial uses, using a desuperheater to extract heat directly from the refrigerant line, are particularly appropriate for restaurants and other facilities that have a high hot water usage as well as high peak demand.

The heat pump/water heater combination is substantially more efficient than even a heat pump with a gas water heater. This lowers the electricity use as well as the demand and provides copious quantities of hot water at the same time. Usually there is a large storage tank to retain water for off-peak hours and an auxiliary heat source for use when the heat pump cannot satisfy the demand.

Power Quality

A technologically more sophisticated means to reduce loads and electrical demand profiles is to increase the quality of the power supplied to a facility and its equipment. Poor power quality has several implications in the overall electrical distribution analysis. Power quality is synonymous with usefulness. A three-phase ac service trashed by many harmonic waveforms is not used as easily or efficiently as a clean three-phase sinusoidal waveform. The more harmonic distortion in a system, the greater the percentage of unusable power that is delivered from a source.

Harmonics are a difficult concept for most electrical designers. Mathematically the term *harmonic* refers to multiples of a specific fundamental waveform. The second harmonic has twice the frequency of the fundamental harmonic, the third harmonic three times the frequency, and so forth. When a power source is infected by harmonics, the frequency of the source is not clean and consistent. A current at just twice the nominal frequency, for example, will try to rotate a motor at twice its rated speed. This impulse may happen only randomly, but when it does, it introduces spurious dynamic forces in the motor. The effect can be even more pronounced in sensitive electronic equipment, such as a motor turning a computer hard disk drive. Even a small variation in speed can have disastrous consequences to the storage and retrieval of data from media that depend on accuracy down to the size of molecules and atoms.

Harmonics are thus unwanted variations of the usual voltage, frequency, and amplitude of the waveforms moving through a circuit. Electrical equipment is designed to operate when the windings are energized by a specific waveform and phase distribution. When this is not matched by the power source, the device is not able to convert all of the power delivered to it to do actual work. Three equal sinusoidal patterns of equal amplitude and frequency do not actuate, for example, a motor in a uniform transfer of power. Instead, odd phase angles, disparate amplitudes, and unusual frequencies characterize the service. Some patterns may even be acting to oppose the standard waveforms, trying to turn the motor in the opposite direction from the prevailing currents. Thus, not only do harmonics render the effect of the sinusoidal waveforms unusable, but they can even act against synchronized waves, thus making them unusable as well. The power is all lost to useless heat in the coils and windings.

Harmonic patterns in an electrical circuit are not rare, random circumstances that can be ignored. Nonlinear loads include many common appliances in a facility:

- Static power converters
- Buck-boost transformers
- Fluorescent and other arc-discharge devices
- Resistance welders
- Magnetic cores that need third-harmonic current, such as those in transformers and rotating machines
- Synchronous machines, which create fifth and seventh harmonics
- Adjustable-speed drives for air handler fans, cooling tower motors, pumps, and other hvac equipment
- Solid-state control devices to modulate lights, heaters, and other devices
- Switched-mode power supplies common in personal computers and many industrial instrumentation packages

- Electronic lighting ballasts
- dc-to-ac inverters

You can see that harmonics are more the rule than the exception. Almost every device with solid-state, silicon-based microprocessors generates harmonics. Increasingly, almost every appliance on the market is controlled by electronic means, even such basic devices as toasters and washing machines.

Symptoms

Harmonics are high-frequency, low-amplitude waves that are created in power circuits by nonlinear loads. Consider a three-phase system with a common neutral. When the loads are all linear, the waveforms remain sinusoidal and there is only a small neutral current that is a function of the imbalance between phases. When harmonics exist in a system, there is a neutral current even if the three phases are perfectly balanced. The harmonic waves are not used or balanced or canceled out; they remain in the circuit and are grounded to the neutral conductor. The results of this high neutral current are overloaded, overheated neutral conductors and transformers. Sometimes the distorted waveforms even cause fuses and circuit breakers to trip at far below their rated capacity.

Another effect is the distortion of the standard sinusoidal waves carrying power in the circuit. It is the uniform relation between voltage and current with time and between the three phases that makes a motor run evenly because of regular excitement of the static coils. When the waveforms are distorted, the transfer of power to the motor is erratic, and stresses develop that hinder performance, reduce efficiency, and cause overheating in the motor windings and power conductors.

The most common remedy for harmonic interference is just to derate transformers so that they can handle the extra neutral current. This solves the overheating problem in most cases but does nothing to ameliorate the problem of harmonic distortions in the system.

Power Conditioning

Because it is inevitable that harmonics will exist in the power distribution system of any modern facility, the objective of the electrical designer is to minimize their impact on the overall quality and efficiency of the system. The key to accomplishing this task is the transformer because it is the least-expensive device to condition power. Furthermore it can be used by the clever designer to achieve cleaner power without alerting the owner to the dangers posed, which all too often opens the door to the salesperson marketing ultrasophisticated devices that cater to the owners' fears of the elusive but sinister harmonics. Granted, there are quality devices that get the job done, but for small relatively unsophisticated installations they are excessive for the needs of the typical client.

It is a safe assumption that if harmonics are created in one part of a circuit or larger power distribution system, they will eventually migrate to all parts, unless actively blocked from doing so. In the case of individual circuits, a dedicated circuit can be run to each point of use. Circuits from a single panelboard are not much different. All the circuits are still wired in parallel and are no less connected than individual duplex outlets on a single power circuit. So, even if computers are only used on the first story of a three-story building, their damaging harmonics can turn up anywhere. If no computers are used anywhere, the electronic fluorescent lamp ballasts can still trash power quality. If computers and electronic ballasts are both used in a building, the power quality throughout the building might even be worse than if either were used alone, and the poor quality will be evidenced everywhere.

Other than the panel busbars, the main culprit for spreading harmonics from one circuit to another is the close proximity of circuits in wireways, cable trays, and feeders. The induction between the wires is sufficient to induce harmonics in circuits that are otherwise clean. One solution is to install home runs in dedicated circuits, which would be very expensive. The other solution is to design circuits carefully so that (1) each circuit is either a clean circuit or a harmonic one, the twain never mixing, and (2) only like circuits are grouped in feeders or wireways, all clean or all harmonic. The same philosophy applies to the next step up the power distribution ladder, the panelboards.

The key to avoiding harmonic chaos from the source itself is to divide and conquer. Some organizations already require computers to be on separate panels from linear loads. This also makes it possible to provide them with emergency power, as a future alternative. Other than complete isolation of the two systems, an effort should be devoted to minimize the circuits requiring the highest-quality power because this is most costly to create. The power-quality hierarchy, in order of increasing quality needs, is approximately as follows:

- Lighting
- Major elevator and hvac motors
- Building safety, fire, and environmental controls
- Mainframe computer and network servers

Most mainframe computers have uninterruptible power supplies (UPSs). Do not confuse this with isolated power supply. The UPS, although it provides clean power, does so only for a short power outage. The rest of the time the system depends on the building power supply only.

Lighting, power, and equipment systems will already be separated into different services, especially for larger systems and for health care facilities where it is a code requirement. The difference is that, when possible, lighting is often supplied at 277 V and motors at 480 V/three phase. This minimizes the size of feeders, conduit, disconnects, and other installation items. If these systems are to be physically separated by transformers, lower-voltage devices will be used, either 208 or 120 V. This is not necessarily either a very costly alternative or an unreasonable one.

When a lower voltage is used at a facility, fewer devices are installed on each circuit. This means the potential for overall harmonic distortion is reduced because fewer items are contributing. That is, there will be more panels and each panel will isolate a smaller segment of the system from the rest of the power grid. Extra power panels will also be required, because they will be located near concentrations of equipment such as in central plants or elevator rooms. These smaller power panels will, again, isolate their equipment from the whole. The same philosophy applies to the panel service to duplex and small appliance outlets, which is usually 120 V anyway.

The only requirement is that the main building service be 480 V. Other than a direct feed to a larger chiller (which should have its own power conditioning equipment so it doesn't scramble the entire power supply), the voltage is distributed at 480 V to each transformer and then stepped down to each major equipment or branch panel.

Transformers

There are several types of transformers that can be specified, depending on the characteristics of the power source. The idea is to minimize mutual inductance by physically separating the source from the load. This can only be done by isolation transformers, which have no direct connection between the primary and secondary windings. This reduces the passage of noise and transients from the load to the power source. Depending on the sophistication of the device, they also retard the propagation of harmonics.

K-factor transformers are best used for circuits with no expected harmonic distortions. They are regular dry-type transformers that can still handle nonlinear loads because they are designed to handle extra-high temperature rises. Not knowing what new equipment will be used in the future or its harmonic rating, it is best to design for every possibility.

Regular isolation transformers should be used for all known harmonic circuits. If the load can be determined accurately—such as only lighting, only computers, or only large motors with VFDs—a transformer can be specified to block the expected harmonics. The more sophisticated delta-wye isolation transformers have electrostatic shields between primary and secondary windings to isolate the loads from any common load noise that may exist in the power source.

Phase shifting transformers are more sophisticated yet. They use phase shifting methods to cancel harmonic currents. When two harmonic currents of the same magnitude are shifted by half a wavelength and combined, they cancel one another (which eliminates their contribution to useful power of the circuit, but at least it cleans up the source). Split secondary transformers are used to eliminate the harmonics created by VFDs, by a 30° phase shift that cancels fifth and seventh harmonics in 60-Hz systems.

Power Producers

Special filters are a solution to some large individual equipment loads. They are usually designed to counter a specific harmonic that is characteristic for the device. The filter is designed by the manufacturer and is integral to the equipment. Whenever any major device can be purchased with such filtering capability, it should be.

Motor generator sets are another form of filter. They mechanically isolate the load by providing an independent source that has a high quality of power. A larger facility that can benefit from peak load reduction by such a device should also consider the benefits of having a clean, independent power source. In addition to generous rebates for peak load reduction, many municipalities also permit selling power back to the public grid, which can enhance the economics of such an installation considerably because it allows payback calculations to be done using a consistent, dependable loading.

There are several reputable manufacturers that market packaged generator sets that can be used for peak load reduction. The ideal installation has several small generators staged to match the building load. They can be computer sequenced and interfaced with the building automation system not only for maximum reduction in the electricity usage but also to rotate the leading device. Figure 15.1 shows one such installation.

This particular installation is supplied from natural gas mains, which is a clean power source for industrial applications. The devices are self-contained, with individual computer controls to coordinate the gas engine and the electrical generator, for consistent voltage and frequency. The windings are excited by the existing building power

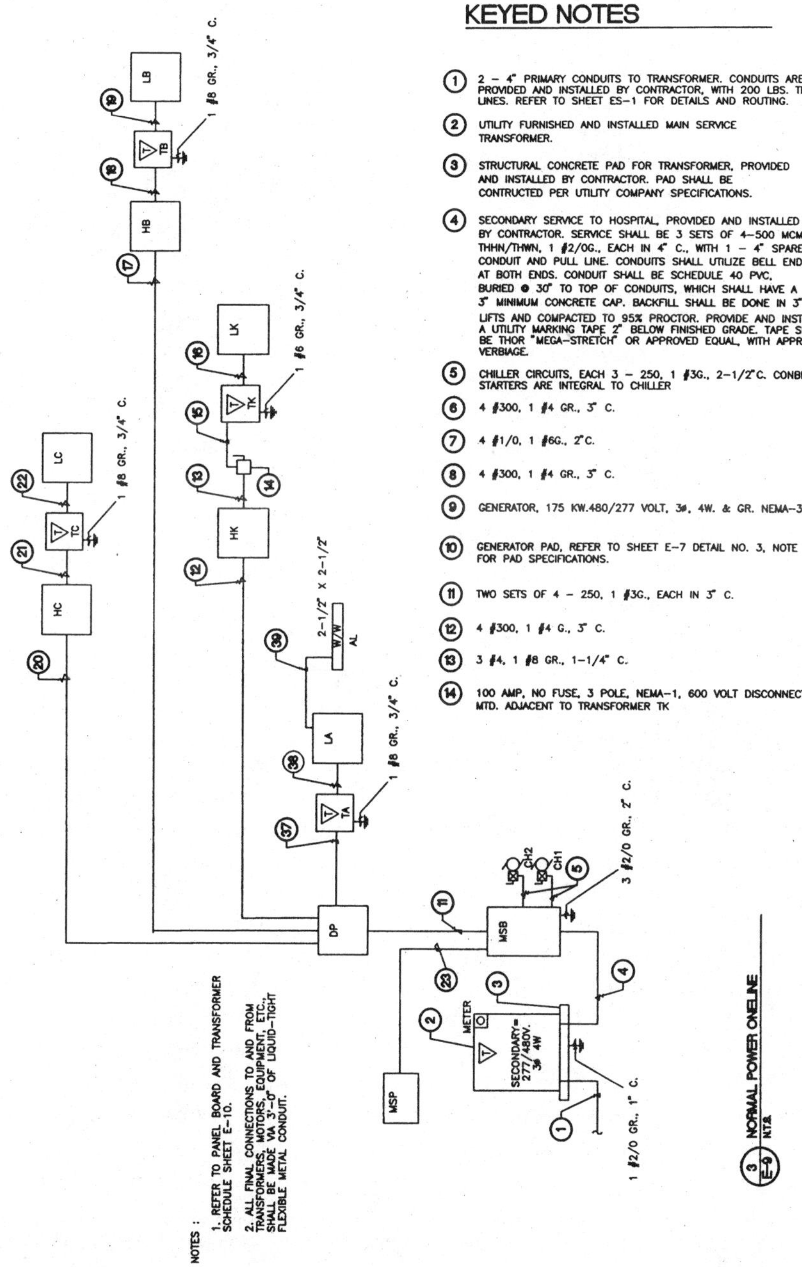

KEYED NOTES

1. 2 – 4" PRIMARY CONDUITS TO TRANSFORMER. CONDUITS ARE PROVIDED AND INSTALLED BY CONTRACTOR, WITH 200 LBS. TEST LINES. REFER TO SHEET ES-1 FOR DETAILS AND ROUTING.

2. UTILITY FURNISHED AND INSTALLED MAIN SERVICE TRANSFORMER.

3. STRUCTURAL CONCRETE PAD FOR TRANSFORMER, PROVIDED AND INSTALLED BY CONTRACTOR. PAD SHALL BE CONTRUCTED PER UTILITY COMPANY SPECIFICATIONS.

4. SECONDARY SERVICE TO HOSPITAL, PROVIDED AND INSTALLED BY CONTRACTOR. SERVICE SHALL BE 3 SETS OF 4-500 MCM THHN/THWN, 1 #2/0G., EACH IN 4" C., WITH 1 – 4" SPARE CONDUIT AND PULL LINE. CONDUITS SHALL UTILIZE BELL ENDS AT BOTH ENDS. CONDUIT SHALL BE SCHEDULE 40 PVC. BURIED @ 30" TO TOP OF CONDUITS, WHICH SHALL HAVE A 3" MINIMUM CONCRETE CAP. BACKFILL SHALL BE DONE IN 3" LIFTS AND COMPACTED TO 95% PROCTOR. PROVIDE AND INSTALL A UTILITY MARKING TAPE 2" BELOW FINISHED GRADE. TAPE SHALL BE THOR "MEGA-STRETCH" OR APPROVED EQUAL, WITH APPROPRIATE VERBIAGE.

5. CHILLER CIRCUITS, EACH 3 – 250, 1 #3G., 2-1/2"C. CONBINATION STARTERS ARE INTEGRAL TO CHILLER

6. 4 #300, 1 #4 GR., 3" C.

7. 4 #1/0, 1 #6G., 2" C.

8. 4 #300, 1 #4 GR., 3" C.

9. GENERATOR, 175 KW.480/277 VOLT, 3ø, 4W. & GR. NEMA-3R

10. GENERATOR PAD, REFER TO SHEET E-7 DETAIL NO. 3, NOTE 4, FOR PAD SPECIFICATIONS.

11. TWO SETS OF 4 – 250, 1 #3G., EACH IN 3" C.

12. 4 #300, 1 #4 G., 3" C.

13. 3 #4, 1 #8 GR., 1-1/4" C.

14. 100 AMP, NO FUSE, 3 POLE, NEMA-1, 600 VOLT DISCONNECT SWITCH, MTD. ADJACENT TO TRANSFORMER TK

NOTES :

1. REFER TO PANEL BOARD AND TRANSFORMER SCHEDULE SHEET E-10.
2. ALL FINAL CONNECTIONS TO AND FROM TRANSFORMERS, MOTORS, EQUIPMENT, ETC., SHALL BE MADE VIA 3'-0" OF LIQUID-TIGHT FLEXIBLE METAL CONDUIT.

Figure 15.1 Main power one-line.

source so that the phase and voltage can be matched exactly. The sets are in weatherproof enclosures, with adequate noise suppression for use in public places, such as a parking lot or parking garage.

Synchronous Motors

The manufacturers of synchronous machines are pushing a novel use of their power-generating equipment, as an energy-conserving device. The premise is based on the high neutral current that is produced in buildings with a lot of harmonic distortions. Normally this current is grounded at the panel or the transformer and lost. An alternative is to collect this ground current and route it to a central location, where it can be used as the power source for a motor or a motor-generator set.

The simplest use for the power is to drive a single motor, such as a circulating pump motor for the hvac system that operates whenever the building is occupied and power is in use. A simple circuit can augment the power to the synchronous motor when the building neutral power is not available.

An alternative is to use the building ground current to drive a motor-generator set and to either return the power thus created to the building grid or to the utility power grid. Either way, the generator will need some expensive, sophisticated controls for it to parallel with the existing service. These controls match voltage, frequency, and phase angle with the building service. They are the same controls an emergency power generator set uses when it is applied as a peak load reduction device.

The difficulty with paralleling is that if the generator voltage is out of phase with the line voltage, the result is the equivalent of a short circuit. This can damage equipment and will probably take the whole system off-line. Voltage, power, and reverse power must all be closely monitored to maintain the coordination between the systems, or the generator is taken off-line.

Using the neutral power to run a pump motor is much simpler. Hydronic circuits are much more forgiving of variations in power. A backup pump on a VFD can be used to make up the difference in power to the fluid loop to ensure continuous and consistent service.

In either application the synchronous machine can be used to correct the building power factor. When the machine is operating as a motor, the reactive power (i.e., the power factor) can be controlled by modulating the field current to the stator coils. This field current can be used to affect the power factor for the entire building service. In large facilities this can avoid a low power factor penalty. In all facilities it can make the power factor closer to unity, for a more efficient and complete use of the power that is purchased from the utility.

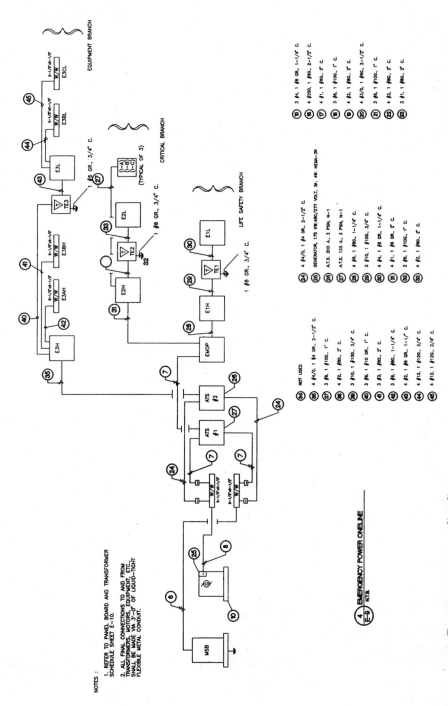

Figure 15.2 Emergency power one-line diagram.

Chapter 16

Special Applications

This section reviews some complex applications of the principles developed in this text, concentrating on the integration of mechanical and electrical systems. The benefits are greatest when both systems are distributed evenly throughout the conditioned space, as opposed to central systems that are the rule for large facilities. The reason to do this is to balance motor and electrical lighting loads as close to the point of use as possible to minimize power factor-related losses. This technique also isolates smaller areas of the facility on local panels. This, as seen previously, has the potential to reduce harmonic distortions if transformers are used at each location. These local panels, in turn, have the benefit of permitting submetering of occupants so that energy conservation can be encouraged.

The water source heat pump technology is featured as the air conditioning and heating system of choice. It is an extremely efficient as well as versatile system. It is available in small or large packaged units and can be used in a number of configurations on the waterside. A common-loop system transfers thermal energy within the building by its very nature, and reheat is never necessary—quite the contrary, heat is transferred effortlessly from other places in the building itself, never requiring the input of an artificial heat source. Finally, a hybrid water source heat pump system with a few ground loops introduces the ability to incorporate thermal storage into the whole system. This allows the use of peak load reduction in all seasons, load shifting, and many other traditional energy management strategies.

Overall, the effect of decentralizing the mechanical and electrical services and using them to take advantage of thermal storage and building mass features is to make the entire facility into a dynamic system. No longer can either mechanical or electrical systems be analyzed as simple linear functions but as the much more complex nonlinear ones. Fortunately, the careful application of the principles established henceforth will make the whole process seamless and relatively easy to implement.

Water Source Heat Pumps

A novel technology is based on using the ground instead of the atmosphere as the heat sink for air conditioning systems. The equipment—which is identical to a conventional DX system with a refrigerant loop, evaporator coil, and expansion valve—circulates water through piping buried in the ground. Heat from the conditioned building is rejected to this fluid in a closed loop, and as it is pumped through the piping, it rejects the heat into the ground.

In moderate climates with a mild winter and a long cooling season it can extract an almost unlimited amount of heat from the ground, at a very efficient rate. The ground acts like a thermal storage system. Thermal energy that is taken out during the day, lowering the ground temperature, is returned at night naturally as the ground near the buried piping returns to ambient temperature. The same loop and air handler can be used for cooling in the summer, which rejects heat into the ground. For a given system, the same amount of heat can be added and rejected on an annual basis, with no net effect on the average ground temperature. If one season is more predominant, the average ground temperature can increase by a degree or two a year, depending upon the imbalance. This is less a problem in colder climates than in warmer ones, however.

There are several aspects of this unique system that make it so efficient. First, heat is rejected to the ground instead of the atmosphere. Because the average temperature of the ground is both warmer in the winter and colder in the summer than ambient, thermal energy can be transferred with a higher rate of efficiency. Second, the efficiency of fluid (refrigerant) to fluid (water) heat transfer is much more efficient than fluid to air heat transfer, so less power is needed to cause the process. Finally, because the operating temperature of the refrigerant is far lower in the ground source system, there is less wear on the compressor, so it lasts much longer and operates more efficiently.

Electrical Advantages

An ultraefficient hvac system has several advantages. Energy use is lower and the power grid is less extensive and expensive to install. The peak demand is lower, not only because the actual energy used by the equipment to provide a given quantity of cooling (or heating) is less but also because of the thermal storage characteristics of the system. The ground within which the piping is buried is part of the thermal system, thereby adding thermal mass to the space conditioning system. Normally only a building with a high mass, such as a concrete or masonry building, exhibits the reduced peak and lower total cooling and heating costs. A properly designed ground source heat pump (GSHP) system can exhibit the same characteristics.

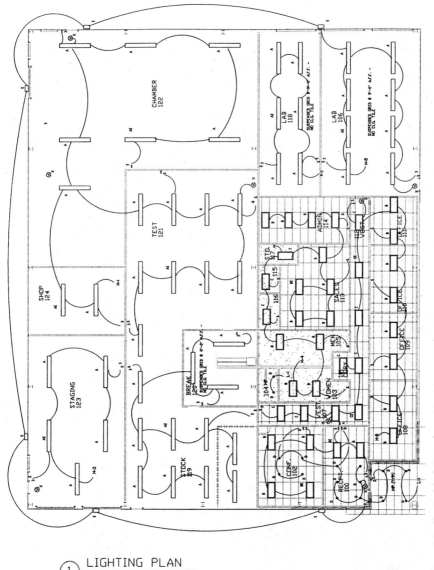

Figure 16.1 Typical lighting plan.

253

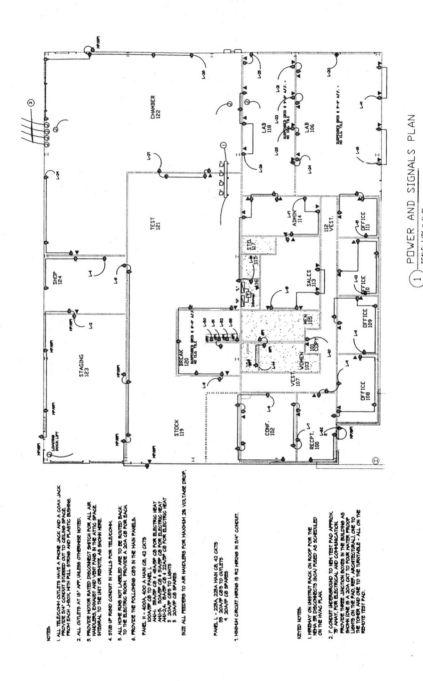

Figure 16.2 Typical power and signals plan.

The greatest advantage happens in areas that have no inexpensive natural gas or fuel oil to fire boilers for space and water heating. This is common in rural and less-developed areas, which must depend upon electricity for heating. The electric resistance heating equipment is not only more expensive to install but can cost up to 10 times more to operate than hydrocarbon-fired systems. The GSHP system is more efficient than even the natural gas or fuel oil-fired boiler system.

The ground source system costs less to install, too. The reason for this is that other systems require a separate furnace, storage tank, and controls, whereas the ground source heat pump can be configured to provide all the heating needs, domestic hot water as well as warm air for space heating. The heat source for everything is the ground.

Replacing the electric resistance heating elements at a sizable facility can reduce the main electrical service by half and the annual utility bills by the same amount, if not more. The ground source heat pump (GSHP) system is not only more efficient in the winter, but it provides efficient cooling in the summer as well. Furthermore, the refrigerant charge of the equipment is substantially less than any other refrigerant-based system, and it is under less pressure, so the equipment is not as susceptible to costly freon leaks into the atmosphere.

Outside Air Systems

High-occupancy buildings are required to introduce large volumes of outside air to keep the space safe. In the cooling season this means the fresh air must be first cooled down below the dew point and then reheated to room temperature. This is done to extract moisture from the air before pumping it into the buildings. A large building can therefore have a very high heating load even in the middle of the summer. This may seem to be more the concern of the hvac engineer than the electrical designer. However, if electric resistance heat is the thermal source for this dehumidifying heat (as is most often the case in large outside air systems), the maximum electricity use occurs at the exact same time as the building hvac load peaks. This can double the peak load and result in inordinately high utility bills.

GSHP systems provide a very efficient, and effective, alternative, one that not only avoids the high power peak when reheat and cooling coincide but also results in a lower peak than conventional systems. The trick is to provide all the cooling with water source heat pumps, configured so that all the units are piped together on a common closed ground loop. The other option is to have a single unit to temper the outside air to the whole building. This is often the case anyway for high-occupancy buildings such as schools and hospitals.

Picture the water circulating in a big loop, with many air handlers drawing fluid from a supply line and sending it back to the ground on a return line. The supply water is cool, and the return is hot. The reheat coil in the outside air system needs hot water, which it taps from the loop and cools it down before returning it to the system. The amount of heat rejected at this reheat coil for a small classroom building can be one-third to one-half of the total building cooling load. Consequently, not only does the system result in almost free heating, but in doing so it provides a substantial quantity of free cooling.

This can be an extremely effective system in hot and humid climates with a high peak cooling load and a lot of humidity in the air that must be extracted.

Hybrid Systems

The most common large industrial system configures the ground piping as vertical drill wells 300 or 400 ft deep, about one well per ton of cooling required. These wells are quite expensive and greatly increase the cost of the installation. It is often more economical to design a hybrid system. In southern climates, the ground loop is sized to handle just the peak heating load in the winter, and a cooling tower is provided in the loop to handle most of the cooling load in the summer. That way far fewer wells must be drilled, but the ground loop is still available for peak load reduction in the summer and for enhanced thermal mass effects all year long.

A common occurrence in such climates is that there will be several days throughout the heating season when the air conditioning must come on. Schools have a high people load, which can heat up a space quickly. (Any building with interior zones, those without any roof or outside wall exposure, will require cooling for these spaces all winter.) A conventional system will either have to shut down the boiler and start up the chiller or run both systems simultaneously. The latter is the most common solution, although either system will be very inefficient.

A water source heat pump-based system will not only operate more efficiently than the aforementioned alternatives, it will operate more efficiently than either the conventional heating or cooling system at its peak efficiency on its own. The water source heat pump can heat or cool off the same common loop, without any change in configuration other than an automatic switchover that the system does by itself. In fact, when some units are heating and some are cooling, the system operates more efficiently than if they are all doing either heating or cooling. This means the water source system is ideally suited for large buildings with a high proportion of interior zones.

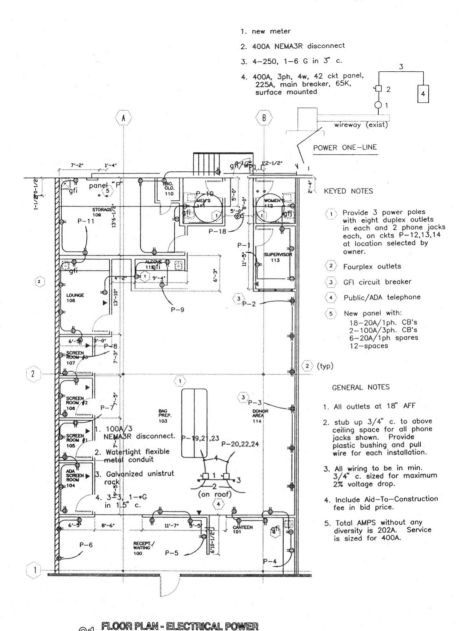

Figure 16.3 Typical clinic power plan.

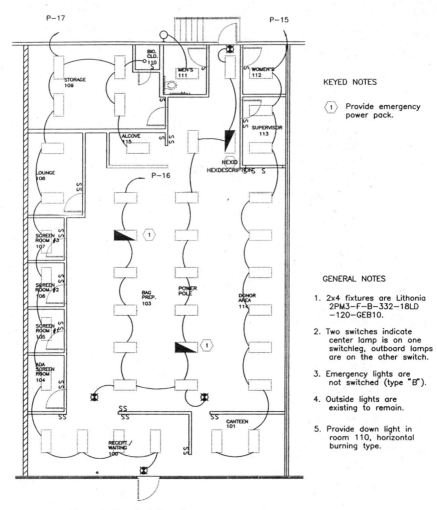

Figure 16.4 Typical clinic lighting plan.

Complex Systems

The idea with more sophisticated systems is to design them to reduce the use of electricity as an auxiliary power source even further and to reduce the electric bills when electricity is used. Although this would normally be the responsibility of the mechanical or hvac engineer, when the difference in power use becomes so potentially large, it behooves the electrical engineer to investigate the potential savings in installation as well as operating costs.

There are several key concepts that make the hybrid systems an attractive alternative:

1. Water source heat pumps can provide heating and cooling with simple, automatic switchover.
2. When many systems are combined on a common hydronic loop, some can be heating and some can be cooling. In fact the more simultaneous the heating and cooling systems, the more efficient the entire system.
3. The same technology has been modified to permit heating of hot water for domestic use, off the main loop.

The latter item is a new idea in this presentation. Before, existing packaged systems were modified with shell-in-tube heat exchangers to heat water from the refrigerant. Now this is replaced by a completely separate system that uses a refrigerant cycle to simply transfer heat from the main loop to the domestic water stream.

There are any number of uses for this water. It can be used in the reheat coils of a large central outside air system or to heat water for use in the kitchen, laundry, showers, or restrooms. Whenever this is done, the heat extracted from the water saves the trouble of having to do so at the cooling tower, chiller, or ground loop. This can be managed effectively to do several things in the overall engineered design of a facility:

1. Reduce the size of the chiller and cooling tower
2. Reduce the size of the boiler
3. Reduce the annual operating hours of all these systems
4. Manage the peak load to reduce demand charge fees

The latter issue is of the most interest to the electrical engineer. No one else on the design team can address this issue with enough authority to force the consideration of important energy- and money-saving applications. There is no incentive for the mechanical engineer to do so because it would reduce the cost of the mechanical systems (and thus the design and construction fees) and would involve more risk in the implementation of unaccustomed control strategies. Although the necessary controls are well known, frequently implemented, and successful in saving energy, they are just new enough for conservative engineers to shy away from them.

To get an idea of exactly how important this issue is, consider that typically half of the electricity saved in large commercial conservation projects is in demand and half in use. Thus, by reducing the peak hvac demand, the savings that result are double. Often it is even more than that because the peak charges in the single worst month also are ratcheted to increase the fees on adjacent months—in some cases, for the whole year. This is not an unreasonable way for the utility to charge because they must still maintain full service and capacity all year.

Balanced Electricity Use

The objective is to reduce the peak air conditioning load, which in turn will reduce the electric bill by a factor of 2 or more. The use of simple controls, with a little creative scheduling of activities by the staff, can do just that. For example, consider that the peak cooling load is from 3 to 6 p.m. during the summer months at a large hospital facility. A high-occupancy building such as this will condition enormous quantities of outside air, which can double the hvac costs and peak load. This is not good, for the reasons already discussed.

The advantage of a water source system is that at the same time as the peak building load occurs, the system can also handle a heating load that offsets all or part of the cooling load. The laundry can schedule the washing machines to run at this time, the kitchen can clean and cook for the evening meal at that time, and patients can be asked to do their personal hygiene during these hours. The result is that the hot water is virtually free, if not better even than that because the air conditioning system operates so much more efficiently.

A large facility can afford an even more elaborate system. Large storage tanks can be installed and used as backup and buffer. When the system needs to be off-loaded, but no load is readily available, the storage tanks can be used as an artificial load. Hot water is created and stored for use at another time or to handle the peak hot water load of the facility. This is common even in gas-fired boiler systems, and the controls are standard with many manufacturers.

Electric Power at Sea: Surface Combatants

It is hard to imagine a more challenging environment for the design of a safe and sturdy electrical system than a ship at sea. Not only is the vessel traversing the nemesis of electricity, water, but the presence of water is ever threatening all systems. Often it is salt water, which is quite corrosive. The vessel must generate its own electricity dependably and efficiently and in an environment that has a high level of continuous vibration. The role of electricity is even more important because the safety and welfare of the crew and cargo cannot be maintained without power: to navigate, communicate, and control all shipboard systems—many of which are sophisticated computerized installations that are very dependent on a steady power supply.

The problem is further complicated on surface combatants. The electrical systems there must withstand far worse vibrations, including battlefield damage, and extremes of heat, cold, and weather. Furthermore, when such a ship is damaged, its vital systems—navigation, propulsion, and weapons systems necessary for defense and above and below sea surveillance—must be maintained, often at the loss of all other functions on board. Finally, all of these conditions must be met at a premium of space and weight and must be capable of being kept in optimum working order by individuals usually of lesser competence than much of the extremely sophisticated equipment under their charge—and by individuals pressured by many other demands of shipboard life, other than just maintaining the electronics.

Keep in mind as various circumstances are presented, the pertinence to less-exotic settings. Large office buildings with a self-contained power and air conditioning system, for example, have many of the same problems—even if less severe. Reducing overall power use can allow emergency-generating capacity to be downsized, lowering the overhead of all tenants. Distributed air conditioning systems, remote from the most common centralized, chiller-based arrangement, allow individual tenants to be more easily billed for air conditioning loads and permits the building owner to offer special services for computer rooms, such as lower temperatures and extended hours. Finally, using more dependable systems will keep tenants cool and happy.

Collateral Issues

Before evaluating the overall electrical system, it is important to minimize the energy used by the air conditioning system, the single largest use of electricity on board. Typically, this is a central, chilled water system comprised of a chiller and heat exchanger, with the ocean itself as the heat sink or "cooling tower." This is usually the most efficient installation, although it has its drawbacks. A problem with the chiller can disable the cooling systems throughout the ship, including loss of power or damage to the chiller, pumps, or main circulating loop. This, in turn, can threaten sensitive electronics on board, affecting communications, navigation, targeting, and other vital functions. It is imperative that the operating temperature of these systems remain cool to avoid overheating, so every effort must be made to design an effective and dependable electrical system.

These flaws inherent to a central air conditioning system are problems in all ocean-going vessels, which are increasingly dependent upon electronic controls for all vital systems. Furthermore, although a warship may be subjected to battle damage that can interrupt the central systems, other ships are only slightly less susceptible. Rough seas, strong winds, loose cargo, or even weakened bulkheads and structural members on vessels nearing the end of their useful life—all of these things have the potential to cause as much damage as any weapon strike. As profit margins narrow, fuel costs increase, and as competition heightens, mechanical systems will be less well maintained and will be potential trouble spots, such as ruptured, rusted pipes and fittings. From this perspective, commercial vessels will probably experience more problems than combatants, which have more funds and personnel to devote to maintenance and upkeep and less service on the open sea, not being driven by time and tide to deliver goods quickly and in good order.

HVAC Systems

The flaws inherent in a centralized hvac system can be avoided by localized, point-of-use systems comprised of WSHP units. These are self-contained heating and cooling packaged units that are rugged, dependable, and—collectively—just as efficient as a central chiller system. Other distinct advantages include (1) lower maintenance, (2) lower installation cost, (3) less space use, (4) lower freon charge, (5) better humidity control, and (6) greater total cooling capacity.

The most significant aspect of a WSHP-based system is that it uses less power to operate, with a very distinct advantage at partial load conditions—at which chillers are notoriously inefficient. Collectively, the WSHP system provides a much greater cooling capacity for the power it consumes.

WSHP air handlers are more easily integrated into a central energy management system (EMS) than air handlers on a central chiller because the local control capabilities are greater. The WSHP system can even be configured to use less power under way. This has important consequences in sizing the total electrical capacity of the vessel—which has even more of a need to reduce peak load than land systems—thereby enabling downsizing of the generator size, the fuel capacity needed on board, and even the entire electrical distribution system.

The Equipment

A WSHP is an air-handler-sized device with a compressor and a water-to-refrigerant heat exchanger. Water is pumped through the heat exchanger by a small pump, and the refrigerant discharges its heat to the water. The system can cool (at rated capacity) at water temperatures up to 105°F, and heat at water temperatures as low as 40°F. The cooling capacity throughout this temperature range is relatively constant, which is important because it is typically the most predominant electrical load and hvac design consideration.

A WSHP at sea, circulating sea water through the system, operates against an unlimited supply of cold water. This allows the heat pump to operate at above design capacity at all loads. The seawater can be circulated directly through the heat exchanger, if a quality strainer keeps solid particles out of the fluid loop.

A typical WHSP has a small pump to move approximately 3 gallons per minute (gpm) per ton of cooling capacity through the unit. Water is drawn directly from the sea at the hull, below the waterline, through a short length of special PVC piping. When under way, the cooling water intake and discharge fittings can be configured to move water through the fluid loop by virtue of fluid pressure differentials, so the small in-line pump does not have to operate. This reduces energy use by about 20 percent, at a time when it is important to minimize electric power consumption—at sea, when the extra weight and space of fuel and generators can be put to more profitable use carrying other supplies or cargo.

The system does not require a chiller, boiler, pump, or a hydronic loop, as in a chiller-based system. In such a central system, a break in the hydronic loop disables the entire hvac system and threatens the operation of delicate electronic equipment. A WSHP system, on the other hand, is made up of individual, independent systems. Each serves a single zone. If power or fluid circulation to one WSHP is lost, no others are affected in any way.

Application of the Technology

Humidity is always a big problem at sea. The salty environment and the extreme sensitivity of electronic equipment to both humidity and salt exacerbate the situation. Humidity control with a DX system is almost impossible and is difficult with a chiller-based system unless the complexity is increased many fold by adding VAV box(s) with electric reheat.

A WSHP system has superior humidity control with no modifications other than reducing the fan speed or by specifying a variable-speed fan motor controlled by a space humidistat. Both of these features are standard stock items that add little to cost. Also, because they are the manufacturer's standard accouterments, they are dependable and rugged.

A decentralized hvac system is less suspect to other variables at sea. With each toss and turn of a small ship, the fluid balance in a chiller is disturbed, imposing inefficiencies and extra stress on the vessel, pumps, and piping. This is not a problem with a WSHP, which has a much lower refrigerant charge and which uses the identical configuration of components for vertical and horizontal models. Plus, the extra forces imposed on the total system by changes in spatial orientation are small because the material quantities affected are small.

A common feature is the installation of a desuperheater on the WSHP. This is a simple shell in a tube heat exchanger that extracts heat from the refrigerant loop (the tube side of the bundle) to the domestic water loop (the shell side of the system). This provides free hot water, available at local access throughout the ship. It also enables the WSHP to operate even more efficiently.

Perhaps the best feature of WSHP air handlers is their very low maintenance, due to a large extent because the compressor operates at a relatively low temperature. The Austin (Texas) Independent School District has had over 6000 installed tons of GSHPs in operation for more than 10 years. They have had only a handful of compressor failures. Seaborne WSHP's will operate at even lower temperatures and should be even more dependable. This is a very important feature on surface combatants, because there is so much computer and electronic equipment on board that is vital to the survivability of the mission.

Now that the air conditioning system has been explained, there are two things to keep in mind. First, the power use is less when under way, as is the peak loading on the electrical power plant. Second, the power used by the equipment is distributed evenly throughout the conditioned spaces. Each of these items has important ramifications in the design and flexibility of the electrical power distribution grid.

Electrical Systems at Sea

There are thee important criteria guiding the design of power distribution systems for sea-going naval vessels: space, weight, and efficiency. The power grid itself must occupy a minimum amount of space and weigh as little as possible. The grid must also deliver the electricity generated at the source with a minimum of line losses, for optimum system efficiency. The powered equipment itself must be the most efficient, too.

The specification of heavy equipment is the first task. All motors should be high-efficiency top-of-the-line models. They must be properly sized for the load served. If possible, undersizing slightly will maximize their efficiency, if the loading is only occasionally above the maximum rating. Two-speed motors are called for on all systems with a load that is not constant, especially for motors that operate for long hours. VFDs are considered only when a compelling need exists. Otherwise, the harmonics they create are debilitating.

The next greatest electricity demand is for lighting. Fluorescent fixtures with T8 lamps and high-efficiency, low-distortion electronic ballasts should be standard. Compact fluorescent lamps should be used in lieu of incandescent bulbs whenever possible. Motion or infrared controls are needed in seldom-used areas. Lighting panels should be used in all large spaces, such as the mess hall, for dimming and multilevel switching.

These and other related measures minimize the total capacity required of the system at the fundamental level. Reducing the variance of the system power factor from unity will help even more to reduce power as well as harmonics. The best way to do this is to match the lighting and motor loads on each feeder. A WSHP system, with air handlers—instead of a central chiller system—distributes the motor load evenly throughout the vessel. Small panels uniformly distributed throughout the ship will have equal motor and lighting loads, reducing the neutral current lost in the feeders. The other possibility is to have a capacitor bank at the electric chiller motor and at all other large motor loads on board.

Distributed Power Control

An advantage of a power riser that has many small remote panels arranged in an even pattern throughout the ship is that there is less conduit and fewer feeders. This saves money and space as well as power loss through voltage drop in the feeders. The loss, in combat or through shorting or other circumstances, will not isolate more than a small portion of the ship from power.

Perhaps the best aspect of a distributed power system like this is that the individual panels can be "submetered." This is one of the most effective ways to reduce energy use under any circumstances, by holding individual "tenants" or departments responsible. With a decentralized WSHP system, the accountability extends to the hvac energy use, a significant percentage of the total electrical load. The electricity use and demand at each meter can be accurately monitored by a current loop, which can be incorporated into the energy management system.

Another possibility is to distribute the power to the remote panels at high voltage. This reduces feeder size, resistance losses, and space requirements. The transformers, which will be hot all the time, will experience significant standby power losses at all low-loading conditions. Having several small panels/transformers instead of a few large ones allows the power to the small ones to be off completely at times if they are matched carefully to the space occupancy use.

An option worth considering is to distribute all power at the local, duplex outlet level at 220 V. This will reduce the number of home runs, with a resultant savings in weight and space (as well as losses due to voltage drop), and it also allows lights, motors, and other equipment to operate at higher efficiency.

Harmonic Power Distortions

Computerized equipment is used throughout a surface combatant. Many small panels, each with its own shielded isolation transformer with an oversized neutral, can minimize harmonic distortions. Because few harmonic distortions can move back through the circuit past the transformer, there is small opportunity for large distortions to occur in the rest of the power grid, causing trouble to important and sensitive components. The distortions can potentially be limited to the size of each panel rating, and having synchronous motors on each circuit to bring the power factor to unity further reduces the harmonic distortions within the circuits.

A neutral current will still occur for the system as a whole, whatever the preventative measures taken at each feeder. A synchronous motor can transform this current into useful work that might otherwise be wasted. This boosts the overall system efficiency and reduces the peak loading and the required generating capacity.

Communications Backbone

Optic cables should be used instead of shielded coaxial or twisted pairs to channel all data and signals throughout the ship. This reduces space and weight of the communications grid. Also, power requirements will be less and data transfer will be faster and with lower transmission losses.

Central Energy Management System

A full direct digital control (DDC) based EMS is needed not only to minimize energy used for hvac and lighting but also the peak electrical load, by load shifting and load shedding. Reducing the peak reduces the peak generating capacity needed, while reducing the space conditioning and lighting loads and lowering the necessary fuel-carrying capacity of the vessel.

Oil Field Applications

A typical drilling or workover operation—prospecting for new reserves or refurbishing an existing production well—is in a harsh, remote location. The rig generates all of its own power, which can be a significant quantity. The machinery needed to rotate and pick up a couple of miles of steel pipe, against the friction of the wellbore and other mitigating forces, can be great.

Often the rotary table and the turnbuckle, used to rotate and pick up the pipe, respectively, are powered directly by diesel engines, so there are no electric motors used for these operations. The pumps that circulate the fluid in the wellbore are also driven by diesel engines. These are the high-torque high-horsepower demands at the drilling rig, and direct use of fuel is the most effective way to produce this power.

The rest of the equipment on the drilling rig, barge, ship, or platform is typically all electric. This includes the kitchen, air conditioning, lighting, and auxiliary controls for the drilling operation. A drill barge, ship, or offshore platform can have up to a hundred people on board, working two or three shifts 24 h a day. This keeps the services operating at a high level constantly, and the total energy consumption is correspondingly many tens of thousands of dollars a month.

It is common for energy to be used in such situations without regard to the cost. The work is hard and dangerous, and the few small creature comforts such as very cold quarters in the summer and hot food cooked any hour of the day are small compensation for the high stress level experienced by all personnel. This cavalier attitude can greatly increase the operating costs of the rig, however. These costs not only reduce the profitability of the company but reduce the pay of the personnel on site.

Because all of the power used must first be generated from fuel oil, an across-the-board program to economize is perhaps the most prudent approach.

Hot Water Usage

There are several nonoperational functions that have a high power use for short periods throughout the day. These functions include operation of the laundry, kitchen, and use of hot water for personal hygiene. Generating the hot water used in these functions by oil-fired boilers is the best way to reduce the power use. Even more economical is to use the drilling or production fluid, often quite hot as it leaves the ground, as the heat source. Drawing extra thermal energy from this fluid, by a simple shell-in-tube heat exchanger, offers a virtually unlimited source of heat when fluid is circulating. It also reduces the temperature of the fluid, increasing the useful lifetime of the drill bit, piping, and other downhole accouterments as well as saving costs on the treatment of the drilling fluid itself. Fewer chemicals will be needed to maintain the viscosity and other characteristics of the fluid, which means less disposal costs, less environmental impact, and a more stable and dependable circulating fluid (on the basis that the simpler the composition, the more predictable its performance).

The technology to take advantage of even small fluid temperature differences is well developed, in the geothermal applications field. A common system circulates fluid into a deep, hot fractured formation that transfers thermal energy to the fluid so that it can be used to operate a power cycle on the surface to generate electricity. One company retrofits old navy turbines for this purpose, building a self-contained skid with all the necessary pumps, valves, and controls preengineered to provide ready power at even the most remote site. Other systems have been developed with ammonia and other working fluids to take advantage of temperature differences of 100°F or less. Most have been proven quite effective to extract power consistently and efficiently from geothermal fields around the world.

A far more developed appliance is the WSHP. This system operates on principles quite like a direct expansion heat pump but uses water as its heat source/sink instead of the air. This technology can be used directly to extract heat (for space conditioning) from the circulating fluid and with a factory attachment can also be used to heat water for domestic use. It can also discharge heat to the operating fluid, up to temperatures of 105°F, for air conditioning.

Either of these systems—the industrial-sized geothermal engine or the commercial-sized heat pump/water heater—can be used to diminish the need for an artificially generated heat source at remote sites. This not only reduces the required capacity of the electrical service but also the fuel storage capacity and the initial cost of the installation.

Peak Demand Reduction

Reducing peak demand concepts may seem out of place in this circumstance. However, many of the motives that influence large power companies to reduce peak demand apply here as well. A high peak causes the need for an expensive and bulky system to generate, transform, and deliver the power. Furthermore, because transformers, motors, and other equipment must be sized for the highest capacity, it operates less efficiently at all times other than peak. This further contributes to the inefficiency of the overall system.

Chapter

17

Designing in CAD Programs

There are many computer-drafting programs on the market. They range in value from a $12 basic on CD to a $10,000 monster. These programs are appropriate for a variety of projects, from a simple residence to a multistory hospital, respectively. The really fancy programs have a basic program plus advanced modules with extensive symbol libraries, automated design functions, and three-dimensional graphing capabilities. Most of the fancy routines are for architects and mechanical designers. Electrical design uses so few symbols that a full library of symbols takes only a few hours to create from scratch.

On the basis of a modest-sized facility, you cannot do any better than AutoCAD LT. This is a stripped-down version of the full AutoCAD software. The latest version is compatible with Windows 95 and is actually much faster than the full version of AutoCAD because some of the cumbersome routines are not included. The program is smaller too, so it takes up less space on the hard drive. Otherwise it is identical in every way to AutoCAD 14 for Windows, with the exception that it cannot run custom AutoLisp routines. These are fancy software routines that designers and operators are convinced save hundreds of hours of drafting and design time. Most managers are not familiar enough with AutoCAD to realize that the truth of the matter is that those little custom routines waste, rather than save, hundreds of hours of design and drafting time. If the store-bought package does not have the routine, it is not worth having. In fact, it is a distinct advantage to standardize on the AutoCAD LT for the simple reason that it prohibits operators from doing any customization whatsoever.

Customization is the archenemy of productivity in a CAD operation. In an office with more than one workstation, it is inevitable that every software package will be completely different. This means several things:

- When the primary operator is out, no one else can do any work at all on his or her computer because all of the settings and keys are programmed to the operator's specific way of doing work.
- When a drawing is created on one machine, it cannot be easily read, edited, or plotted from another machine that has different custom settings.
- A perfectly basic operation on one machine may crash a whole drawing on a machine that is customized by a sophisticated operator.

This is a very real problem in every engineering office, and it deserves very serious attention by management. The easiest solution is to standardize on AutoCAD LT and then establish specific standards for design and drafting work in that platform alone. These standards should be specific in three areas: external referencing, layer names, and line types. The rules in these three areas should be carved in stone before the first drawing is even contemplated. Otherwise there will be nothing but trouble until the rules are formalized and implemented officewide. They are important even if just one workstation is set up for CAD because they standardize drawings so that editing, copying, and printing are easier.

External Referencing

The first task is to create a suitable title block and border for all the drawings that will be issued by the engineering and design staff. This has the company logo, firm name and address, project name and address, date, revision number, and page number. This should be created in three sizes: 8.5 × 11 in, D-size, and E-size sheets. Each should be stored as a separate file in a separate project directory.

When a new project is started, say a small one, the 24-in × 36-in title block will be opened and edited to the specific project name, date, and other particulars. This will then be promptly stored as a separate drawing file in the new project directory. This is done by work blocking the title block, and it is stored as a separate drawing called titlblck.dwg with the insertion point at the lower left-hand corner of the border. This drawing will be used as a background layer for all future work on the project, just as all drafting work used to be done on title block paper.

The first drawing that is created is the background. Usually the architect will provide the drawing file. If this is the case, the drawing should be copied into the project directory, given a specific name, and stored. Then all of the extraneous information is either turned off by layer or erased altogether (this makes the drawing smaller and easier to regenerate when working on subsequent drawings). When all that remains are the walls, windows, door swings, and other pertinent architectural data, the drawing is saved in the project directory as a work block called backgrnd.dwg. Again, the insertion point should be designated as the lower left-hand corner of the drawing. When the drawing is saved, it should be more or less centered in the screen so that the lower left-hand corner will coincide approximately with the same point on the previously saved title block.

Now it is time to create drawing E1. First a new drawing is opened and named appropriately (usually as projectno-E1.dwg or 9801-E1.dwg). Then the title block and background drawings are externally referenced with the XREF command, with 0,0,0 as the insertion point on the screen. This will align the background on the title block, and both drawing files will be unchangeable from within the E1 drawing file. The only way to change either the background or the title block is to actually open the proper drawing file and edit it there.

The advantage of having this kind of layered setup is manifest in projects with many drawings. Whenever the architect changes the background, it can be updated immediately on all of the drawings by just modifying the backgrnd.dwg—likewise with the title block. Each time new drawing sets are issued, the entire set can be instantly corrected for date, drawing phase, or revision number. This not only makes it simple to change the background, but it also ensures that the background and title block for the entire set are consistent and identical.

Another advantage is that cross-referencing the common parts of the drawings means that the size of each drawing file is much smaller. For example, if the background is 2 megabytes (Mbytes) for a set of 10 drawings, 20 Mbytes of storage are required on the hard drive or file server. Only 2 Mbytes are required if all of the drawings cross-reference the same background drawing. For any given project this may not seem like such a big deal, but after 10 or 15 projects are completed on CAD, you will be amazed at how little free disk space is left on the file server. Then you will start economizing at every opportunity.

Layer Names

Layers are like overlays on mylar when the room names used to be on one layer. When wanted, they can be added to the background layer to print the names on the background. It is not really necessary to put

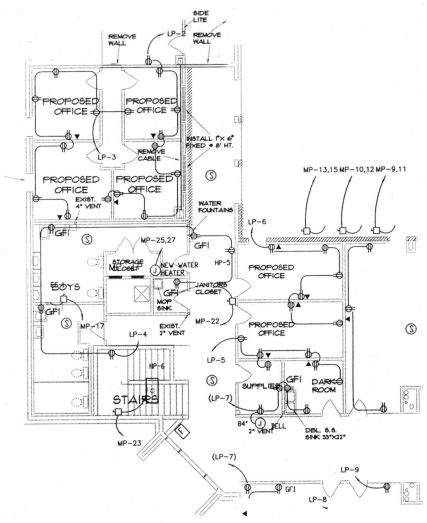

Figure 17.1 Office power plan.

different design elements on separate layers, but it is a good practice to get accustomed to doing so, even on smaller projects. The more complex projects will require layers, and if the system is not learned by first practicing on smaller projects, it can be a far more time consuming and frustrating experience.

The principal benefit of layers for small projects is to separate the disciplines and also to separate elements of a project so that some can be printed out darker or thicker, others thinner or even screened. A good beginning color scheme, using the basic AutoCAD colors, is as follows:

Layer name	Colors
E-Demo	Cyan
E-Power	Red
E-Signals	Green
E-Fire	Magenta
E-Lights	Yellow
E-LtgCkts	Blue
E-Text	White (black)
Bckgrnd	Gray scale

It is a good practice to precede each drawing layer with an E to designate it as an electrical sheet. This is especially important if several engineering disciplines are done in an office. This keeps different disciplines from using the same layer name and really fouling up a drawing.

Notice that there are two layers for the lighting, one for the lights and one for the circuits and switches. This is so that the lights can be work blocked out to be included on a drawing for the reflected ceiling plan, showing the ceiling grid only.

Line Types

The AutoCAD LT package has a plethora of line types, as defined by line thickness and line configuration (e.g., dashed, continuous, and so forth). A very presentable drawing can be created by the use of just three line weights and two line types. More line weights are not easily distinguishable and are not worth the effort. The same applies to line types because electrical drawings are typically created with only continuous and dashed lines. The only exception to these rules is the title block, which can be made quite attractive using very heavy lines for the border, different line types, and all the other features of the software package. Otherwise simple is the best rule, especially because the blueprinting process makes most line weights look alike anyway.

Most symbols and drawing work should be done in the same line thickness, similar to what a 0.7-mm pencil creates. Some designers use such a fine line that it does not stand out clearly against the background layer and dims even further on the blueprint. An extra heavy line, as manual drafting 0.5-mm pencils generate, is reserved for outlining major equipment such as panels, lights, motors, transformers, and motor control centers. Mechanical equipment should be lightly hatched instead of outlined in a heavy line, except that integral motor outlines, panels, or disconnects should be drawn heavy. Outlining these significant items makes them stand out boldly on the plans and makes them easier to find in the context of the drawing.

Extra heavy lines, as created by an extra heavy 0.5-mm line, such as can be generated by the pline command, should be used for the feeders between these major elements.

These line types are compared to the traditional mechanical pencil lead sizes because different plotters and printers respond differently to the colors and line weights designated within the AutoCAD LT program. It is best to try color, line weight, and pen type combinations with the actual plotter and printer available and to standardize on the best three line weights that serve the noted purposes.

Detail Drawings

The exception to the above rules is in the creation of smaller scale, usually $\frac{1}{4}$-in = 1-ft 0-in, drawings. These should have the background drawn with the 0.7-mm line weight, all equipment dashed, and all circuit elements drawn in heavy lines. The reason for increasing the relative line weights is that the background needs to be drawn solid rather than screened so that any unusual features can be clearly shown. In the case of mechanical rooms, the nonelectrical equipment should be screened to indicate clearances, pipe routing, and other features important to the location and connection of the electrical service.

The spaces that most often require $\frac{1}{4}$-in-scale plan drawings are

- Electric and telecommunications rooms
- Mechanical equipment rooms
- Electrical service entrance (with an elevation drawing)
- Other spaces that are so cluttered at $\frac{1}{8}$-in scale that all of the necessary information cannot be shown on the plans

Still, it is best not to get too carried away with $\frac{1}{4}$-in-scale drawings. This always introduces potential conflict and misdirection, especially if the main drawing and the $\frac{1}{4}$-in scale are on different sheets. In very complex drawings, a lot of important information can be inserted on the plans in the form of keyed notes. This helps the circuiting remain clear and easy to follow and provides a ready reference for details on specific circuit elements. The following shapes should be used for these keyed notes:

Shape	Descriptive item
Circle	Keyed notes
Square	Wire sizes
Triangle	Revision notes
Hexagon	Equipment notes

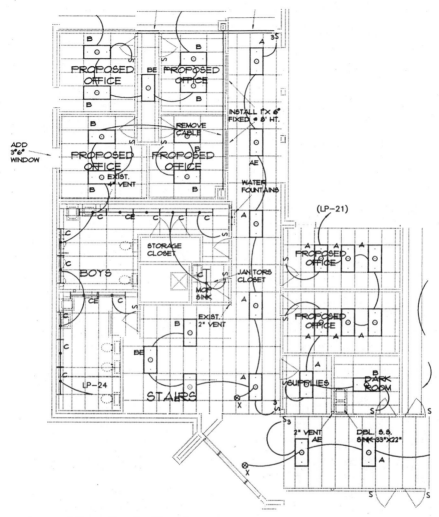

Figure 17.2 Office reflected ceiling plan.

Most text should be ⅛ in high for legibility on reduced plans. A good font is Architext, which is a stylized one that looks much like the old hand-drafted lettering. The advantage of this font is that minor corrections and additions to text can be hand drawn onto the final prints. Architext is close in appearance to a good hand-lettered note. (Always remember to make corresponding corrections to the CAD files after the production crisis is over.) Smaller-scale text is acceptable for such items as installation heights, distances, and other descriptive items. Fixture and equipment marks should be in all capital letters, and equipment mark numbers (not lights) and home run cir-

1. All wiring is 2-#12, 1-#12G in 1/2" conduit, unless otherwise noted.

2. All new outlets at 18" AFF, switches at 42" AFF. Coordinate location of lights, fans, and switches with architectural plans.

3. Replace existing exterior incandescent flood lights with sealed HID lamps and fixtures.

4. All new fluorescent lamps to be 32W energy efficient lamps of the same color characteristic as the existing lamps.

5. If the light level is too high in areas with the new (relocated) 2x4 fixtures, remove one or two lamps.

6. Replace all existing incandescent bulbs with compact fluorescent lamps as they burn out.

Keyed Notes:

1. Install six new recessed downlights in center of the 4x8 panels in the area shown. Use the existing light circuit and switch. Fixtures to be fluorescent, suitable for a damp location.

2. Relocate existing 2x4 fixtures from dining area. Wire to circuit used for existing overhead lights.

3. Relocate four recessed 2x4 fixtures from the existing dining area (select thelights with the newest lenses, lamps, and ballasts), with switches at the door mtd. 42" AFF. Clean lens, lamps, and reflective surfaces. Power from the circuit used by the existing restroom lights.

4. New wall mtd. attic ventilator, switched from the manager's office, and powered from one of the existing AHU circuits (verify circuit capacity).

5. Replace seven existing duplex outlets in the kitchen with GFI devices. Pull grounding wire for each if the existing circuits are two-wire circuits.

6. Replace duplex outlet with a new one (exist. is inoperable) after verifying circuit continuity.

7. New circuit for two new microwaves. Power from an existing kitchen circuit other than those adjacent to this location being used for microwaves.

Equipment notes:

1. Interior recessed down lights (7 EA) to be Lithonia LP6F-26DTT-120-GEB10 with 610 reflector.

2. Exterior recessed down lights (6 EA) in soffit to be Lithonia LS59-2/13DTT-120-BEG10-HPF with C129 reflector.

3. Exterior walkway and front wall lights (8EA) to be Quality Lighting ALP-21-WBA-MH-100-120-BK with remote ballast. Exposed conduit to fixtures to be rigid steel conduit, painted with two coats rustoleum, then with finish paint.

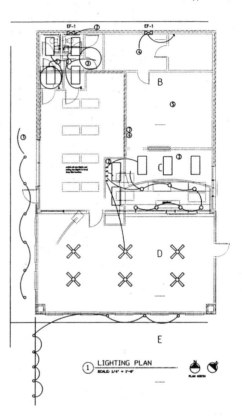

Figure 17.3 Retrofit lighting plan.

cuit numbers should be underlined for clarity. Some common abbreviations for major equipment items are as follows:

Mark #	Equipment
MDP-1	Main distribution panel
MCC-1	Motor control center
XFMR-1	Transformer (or T-1)
H1-1	High-voltage panel
L1-1	Low-voltage panel
AHU-1	Air handling unit
RTU-1	Rooftop unit
CU-1	Condensing unit
VAV-1	Variable-air volume box
VFD-1	Variable-frequency drive
EF-1	Exhaust fan
FACP	Fire alarm control panel
PC	Photo cell
CB	Circuit breaker
ATS	Automatic transfer switch
W/W	Wire way

Another very helpful way to provide important supplementary information is by the use of a General Notes section on the electrical plan. These are extremely important on smaller projects that do not have a specifications section, although any project that is bid (rather than design-build) should have a specifications section, even a nominal one of a few pages.

General Notes

- All electrical work under this contract shall be done in accordance with the NEC, local authority power company, and other special requirements as applicable.
- Final connection to all equipment shall be via minimum 4-ft 0-in liquid-tight flexible metal conduit.
- Electrical contractor shall verify and coordinate the following with the mechanical contractor and comply as required:

    Location of equipment (e.g., motors, thermostats, disconnects, etc.)

    Electrical characteristics (e.g., voltage, horsepower, FLA, control wiring, grounding)

- The contract is to include all charges that may be made to the Owner by the power company, telephone company, etc. All such charges shall be included in the electrical bid for this work. All

278 Chapter Seventeen

General Notes:

1. Electrical contractor to trace all circuits prior to commencing work, and to submit a plan to the engineer for the new circuits if there are important changes from the design documents.

2. All existing conduit, wiring, and circuits to be reused to the extent practicable.

3. All connections to existing circuits must be enclosed in a junction box, per code.

4. All major equipment to be turned on a maximum of twenty minutes apart, to reduce building peak load.

5. City will be running new service to the building in a few weeks. The existing transformer is rated 75 kVa, but the service is only 200A, which is close to the peak building load. City will monitor the service and can upgrade it if necessary. Electrical contractor to coordinate with the City.

Keyed Notes:

1. Re-use existing circuits supplying equipment to be relocated.

2. Several outlets in the kitchen and dining area are in infrequent use. Add new outlets to these circuits.

3. Re-use existing circuit for overhead fluorescent lights.

4. Same circuit as new restroom exhaust fan.

5. Replace circuit to one of the existing condensing units at this location with a new breaker in the panel sized for both condensing units; size wire for maximum 2% voltage drop, and install a small wireway on the wall between the units, with NEMA 3R disconnects to each unit, via watertight flexible metal conduit.

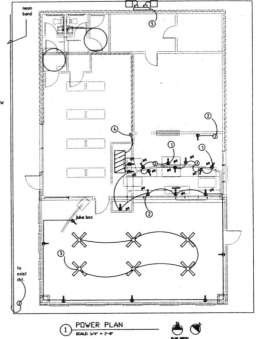

Figure 17.4 Retrofit power plan.

other work necessary for installation of services shall also be included.

- All wiring shall be run in conduit. Thin-wall EMT conduit may be used where concealed above ceiling and in furred spaces, partitions, or walls except where walls are grouted solid. All other conduit plus that exposed on walls shall be galvanized rigid steel conduit. Flexible conduit shall be used for connection of motors. Liquid-tight flex shall be used for outside connections, bonded as per code. Install pull wire in all empty conduits. All conductors are copper. PVC schedule 40 gray conduit is approved for underground installation.

- Grounding of electrical service and equipment shall be as per applicable codes and as indicated on the drawings. Provide ground wire in all flex and PVC conduit per NEC Article 250.

- All conductors shall be copper. Branch riser wiring to be minimum #12 AWG. Other sizes shall be as indicated or as required by code. Minimum conduit size is $3/4$ in.; #14 AWG and smaller may be used

for control wiring when permitted. Insulation THHN for #6 and smaller; XHHW for #4 and larger.
- Device plates in all areas shall be Sta-Kleen. Color of plates and devices to be selected by the Project Architect. Coordinate all outlet locations with all other drawings prior to rough in and notify the Project Engineer of any conflicts. Verify mounting height of all devices with the architect or engineer prior to rough in and comply as required.
- This contract is to include all contingencies that may arise and that may be reasonably required by alteration and demolition work.
- All electrical service to rooftop equipment to be routed through equipment curbs or Thycurbs. No conduit is to be run on the roof.

Of course, the inclusion of such notes does not mean that the designer is excused from preparing a complete and comprehensive set of plans. They are intended only to cover those small things that are overlooked and that a reputable and conscientious contractor would normally take care of at no extra charge.

The best way to deal with the general notes is to create them in AutoCAD once and save the text as a separate file. Then the entire text can be inserted on each project drawing, and edited to the pertinent circumstances.

Chapter 18

Short Specifications

Section 16000

Electrical Specifications

I. General Section

Part 1: General

1.01 Scope of Work

Work covered by this Section consists of furnishing all labor, equipment, supplies, and materials, unless otherwise specified, and in performing all operations necessary for the installation of complete electrical systems as required by these Specifications and as shown on the Drawings. The Work shall include the completion of such details of electrical work recognized as necessary for the successful operation of all electrical systems described on the Drawings or required by these Specifications.

1.02 Work Not Included

Certain labor, materials, or equipment may be furnished under other contracts by the Owner. When such is the case, the extent, source, and description of these items will be as indicated on the Drawings or described in the Specifications. Unless otherwise noted, all labor, materials, and equipment for the complete installation of the electrical work shall be provided under this Section of these Specifications

1.03 Special Requirements

A. Drawings: The drawings indicate the general arrangement of circuits, outlets, locations of motor controllers

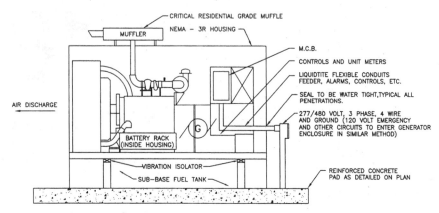

Figure 18.1 Generator detail.

with disconnects, panelboards, conduit routing, and other work. Information shown on drawings is schematic. However, recircuiting or relocating electrical equipment will not be permitted without specific written approval of the Engineer.

B. Shop Drawings: Shop drawings shall include manufacturer's printed information with each item identified as on the Drawings. The information shall include, as a minimum, overall dimensions, weight, phase, voltage ratings, wiring diagrams, and nameplate data.

1.04 Standards and Materials

A. All materials shall conform with the current applicable industry standards, NEMA (National Electrical Manufacturer's Association), ANSI (American National Standards Institute), IPCEA (Insulated Power Cable Engineers Association), IEEE (Institute of Electrical and Electronic Engineers), National Electrical Safety Code, and shall be Underwriters Laboratories listed unless otherwise indicated.

B. Workmanship and neat appearance shall be as important as the electrical and mechanical efficiency. Defective and damaged materials shall be replaced or repaired prior to final interpretations included. The Drawings and Specifications take precedence when they are more stringent than codes, statutes, or ordinances in effect. Applicable codes, standards, ordinances, and statutes take precedence when they are more stringent or conflict with the Drawings and Specifications.

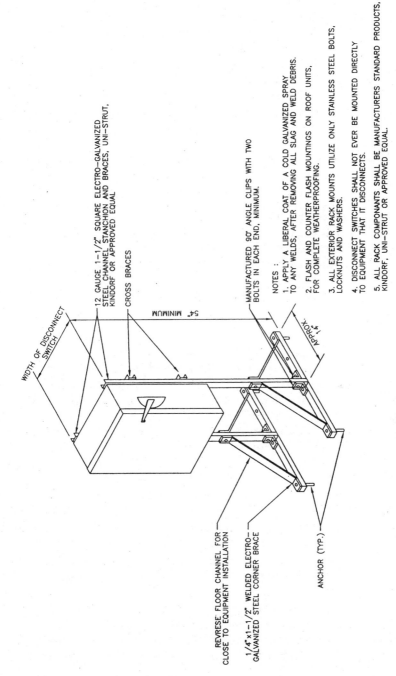

Figure 18.2 Typical disconnect switch.

1.06 Delivery and Storage of Materials
   A. The Contractor shall investigate each space in the building through which equipment must pass to reach its final locations. If necessary, the manufacturer shall be required to ship his or her material in sections, sized to permit passing through such restricted areas in the building.
   B. The Contractor shall retain in his or her possession and shall be responsible for all portable and detachable parts of portions of installations such as fuses, key locks, adapters, blocking clips, and inserts until final completion of work. These parts shall be delivered to the Owner upon completion of the work.

Part 2: Products

2.01 Equipment and Materials
   A. All equipment and materials installed shall be new, unless otherwise specified.
   B. All major equipment components shall have the manufacturer's name, address, model number, and serial number permanently attached in a conspicuous manner.

Part 3: Execution

3.01 Workmanship and Completion of Installation
   A. All specialties must be installed as detailed on the plans. Where details or specific installation specifications are not included herein, approved manufacturer's recommendations shall be followed.
   B. All equipment and material connected with this project shall be installed complete, thoroughly cleaned, and all residue removed from inside surfaces. Exterior surfaces of all material and equipment shall be cleaned and delivered in a perfect, unblemished condition.

3.02 Mechanical Equipment Wiring and Connections
   A. Unless otherwise noted, all wiring for motors, starters, controls, and equipment shall be done under Division 16. All motors for mechanical equipment shall be furnished under other Divisions for wiring as specified under Division 16, except where wired integrally with the equipment. All motors shall operate on 120/208-V, 1-phase, 60-Hz power.
   B. Connection and control diagrams for all mechanical and control equipment shall be furnished under other Divisions

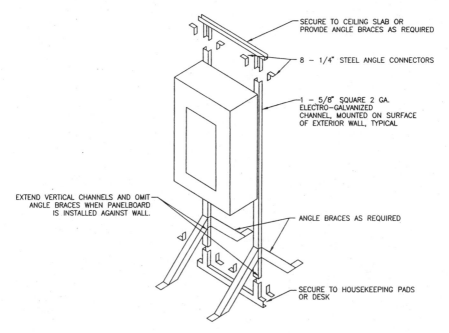

Figure 18.3 Panelboard mounting detail.

and be approved by the Owner for connection under Division 16.

C. Temperature Controls: All temperature control devices shall be furnished and installed under Division 15. Temperature control devices are defined as those which sense or regulate the temperature of flow of water.

D. All line voltage wiring (above 50 V) is specified in Division 16 of these specifications.

3.03 Identification

A. The following items shall be equipped with nameplates:
1. All motors, motor starters, control panels, motor control remote stations.

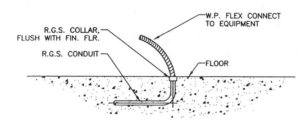

Figure 18.4 Stub-up detail.

    2. All disconnect and safety switches, main distribution panel feeder overcurrent devices and spares, circuit equipment in separate enclosures.
    3. Special electrical systems shall be properly identified at junction and pull boxes.
    4. All branch circuit panel boards shall have identifying engraved plastic nameplates. Also, provide a typed directory card for each branch circuit panelboard. The card is to be placed on the interior side of the panelboard door behind a clear plastic shield. The card shall identify each circuit by number, load, and location.
    5. No Dymo or other stick-on type of tapes will be permitted.
    6. No abbreviations in labeling other than that shown on the Drawings will be permitted without special permission of the Owner.
    7. In general, equipment shall be identified as designated on the electrical drawings. Nameplates for panelboards and switchboards shall include the panel designation, voltage, and phase of the supply. The name of the machine shall be the same as the name used on all motor starters, disconnects, and P.B. station nameplates for that machine.
    8. Nameplates shall be mounted adjacent to motors.
  B. Nameplates shall be fabricated as follows:
    1. Nameplate materials shall consist of 3 ply, $\frac{1}{16}$-in laminated plastic with white core for lettering and black background.
    2. Capital letters shall be used.
    3. Nameplates shall be fastened with cadmium-plated self-taping No. 6 screws $\frac{1}{4}$ in long.

4. The minimum size of all name plates and lettering shall be $\frac{3}{4}$ in high by 2 in long with $\frac{1}{4}$-in letters.

II. Cabinets Section

Part 1: General

1.01 Scope of Work

A. Furnish all cabinets as shown on the Drawings and as herein specified.

B. This section of the Specifications covers all electrical enclosures of more than 150 in^3. For enclosures of 150 in^3 or less see section on outlet boxes and junction boxes.

1.02 Standards

Submittal shall show details of housing, metal thickness, trim, and other considerations.

Part 2: Products

2.01 Construction

A. Boxes and trims shall be made from code gauge steel. Surface-mounted boxes shall be painted to match the trim. Boxes shall be of sufficient size to provide a minimum gutter space of 4 in on all sides, and with ample space for all wires, connections, and equipment.

B. Provide each cabinet with a door and flush catch and lock. All locks shall be keyed alike. Furnish two keys per lock. Doors over 49 in in height shall have a vault handle and a three-point catch complete with lock, arranged to fasten door at top, bottom, and center. Door hinges shall be concealed.

C. Cabinet fronts shall consist of sheet steel panels with a hinged door. Fronts for flush cabinets shall be approximately 3.4 in larger than cabinets on all sides and set so that the front will rest firmly against the finished wall surface.

D. Provide suitable devices for securing, supporting, and adjusting cabinet interiors and fronts.

E. Panels shall not be more than 60 in high.

F. Trims for flush panels shall overlap the box by at least $\frac{3}{4}$ in all around. Surface trims shall have the same width and height as the box.

3.02 Power

A. The exact location of outlets and equipment shall be governed by structural conditions and obstructions, or other equipment items. When necessary, relocate outlets so

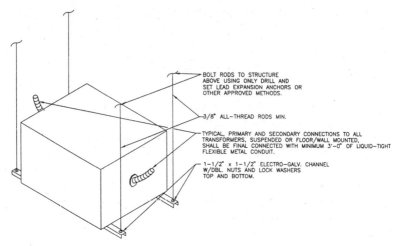

Figure 18.5 Overhead transformer mounting detail.

that when fixtures or other devices are installed, they will be symmetrically located according to the room layout and will not interfere with other work or equipment. Verify any location changes of outlets, panels, equipment, etc., with the Owner.
B. Pressed steel boxes shall be used for interior installations.
C. Structural concrete members and structural steel shall not be drilled or pierced without prior approval of the owner.
D. All boxes shall be mounted so that they are plumb.
E. All boxes shall be mounted so that they are completely rigid without conduit or finished wall support.
F. Back-to-back outlets or through-wall type of boxes are not permitted. Provide 8-in minimum nipple to offset all outlets shown on opposite sides of a common wall to minimize sound transmission.

V. Wire and Cables Section

Part 1: General
1.01 Scope of Work
Provide a complete system of conductors in raceway systems as shown on the Drawings and hereinafter specified. All wire shall be routed through an approved raceway regardless of voltage application.
1.02 Quality Standards

All conductors shall be in accordance with applicable sections of Underwriters Laboratories and IPCEA Standards.

Part 2: Products
2.01 Materials
   A. Wire and cable
      1. All wire shall be copper. Conductors shown on plans are so sized.
      2. All ground conductors shall be copper unless otherwise specified.
      3. Minimum wire size for branch circuits shall be #12 AWG. However, smaller-size wire may be used for control circuits where specified on the Drawings. In no case shall the voltage drop be more than 2 percent.
      4. Copper conductors shall be annealed, 98 percent conductivity soft drawn copper.
      5. All conductors #6 AWG and larger shall be stranded. All power conductors #8 AWG and smaller shall be solid.
   B. Insulation
      1. All conductor insulation types shall be rated for wet and dry locations and shall be in accordance with the National Electric Code for the particular application. All wire and cable shall have the following insulation classes:
         a. All feeders shall be type XHHW, and branch circuits shall be type THHN/THWN.
         b. All insulation shall be rated for operation at 600 V.

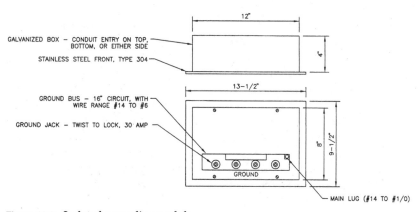

Figure 18.6 Isolated grounding module.

c. All wiring in high-temperature areas, where the temperature may exceed 75°F, shall be rated 105°C minimum. This shall include any wiring within 3 ft horizontally or 10 ft above any boiler or similar heating appliance.

d. All wiring for control systems installed in conjunction with mechanical and miscellaneous equipment shall be color coded in accordance with the wiring diagrams furnished with the equipment. All branch circuit wiring, including circuits to motors, and all feeders, shall be color coded by line or phase and as follows: wire #10 and smaller shall be factory color coded. Wire #8 and larger may be color coded by field painting or color taping of 6 in of exposed ends. Colors shall be provided by the Local Public Utility Convention.

2. Wire pulling lubricant shall be Underwriters Laboratories listed for use with the insulation specified.
3. Wire connections and devices with #8 and larger wire connectors shall be solderless and compression type. Lugs and connectors shall be Underwriters Laboratories listed. No. #10 and smaller wire connectors shall be steel splice caps with nylon insulator.
4. All motor connections shall be the bolted type, with one layer of rubber tape covered with two layers of Underwriters Laboratories-approved electrical tape.

Part 3: Execution

3.01 Installation

Wire sizing noted on Drawings shall extend for the entire length of a circuit unless noted otherwise. Install wire in raceways in strict conformance with the manufacturer's recommendations. Use an approved wire pulling lubricant. Strip insulation so as to avoid nicking of wire.

3.02 Wire Connections and Devices

A. All terminating fittings, connectors, etc., shall be a type suitable for the specific cable furnished. All fittings shall be made up tight. Make up all terminations in strict conformance with manufacturer's recommendations using special washers, nuts, etc., as required.

B. Connect #8 and larger wire to panels and apparatus with properly sized, solderless, or compression lugs or connectors.

C. Connect #10 and smaller wire by using steel splice caps

with proper crimping tool and insulation nylon cap screw over the completed connection. Twist-type connectors will not be permitted.
D. All motor terminations shall be bolted. Bolted connections shall be covered with one layer of rubber tape and two layers of vinyl tape.
E. Lugs and fittings shall be tightened per manufacturer's recommendations.
F. Flash over insulation of joints shall equal that of the conductor insulation.
G. Connectors shall be rated 600 V for general wiring and 1000 V within fixtures.

VI. Branch Circuit Panelboards Section

Part 1: General

1.02 Scope of Work

Furnish and install all panelboards as hereinafter specified and as shown on the drawings.

1.03 Submittals

Submit shop drawings showing details of housing, trim, ratings, arrangement, and type of breakers. Submittal data shall be complete with outline dimensions, descriptive literature, and complete description of the frame size, trip setting, class, and interrupting rating of all breakers and switch and fuse units and panel bus bracing. Available spaces shall be identified.

Part 2: Products

2.01 Construction

A. Interiors: All interiors shall be completely factory assembled with buses and breakers or switches as shown on the Drawings.
   1. Interiors shall be so designed that circuit breakers (or switch and fuse units) can be replaced without disturbing adjacent units and without removing the main bus connectors and shall be so designed that circuits may be changed without machining, drilling, or tapping.
   2. Branch circuits shall be arranged using double-row construction. Branch circuits shall be numbered by the manufacturer.
   3. All spaces shall be fully equipped with bus and mounting straps for the maximum devices that can be fitted into them.

2.02 Panel Designation
  A. Point from which it is fed
    1. Individual branch circuit identification numbers (as shown on the panel schedules) are identified on the plans as to load served and location.
    2. All panelboards shall have a grounding terminal pad for the equipment grounding system. Grounding terminal pad shall be separate from the insulated neutral bus.
    3. The entire assembly shall be thoroughly cleaned inside and out and all surfaces phosphatized and primed with a conductive zinc coating. Inaccessible surfaces shall be phosphatized and primed before welding. All welds shall be ground and sanded to remove the scale formed during welding. All internal welds, scans, joints, and splices shall be wire brushed. Next, all surfaces shall be coated with a primer of zinc chromate, iron-oxide, or as approved. All surfaces shall then be finish-coated with an epoxy resin enamel to an average thickness of 6 mil. Color shall be of a type to which field-applied paint will adhere.
    4. Panelboards shall have general-purpose enclosures and shall be surface mounted except where shown otherwise.
    5. The nameplate to be furnished by the manufacturer, affixed to the dead front of the panelboard, shall contain the following information:
       a. Manufacturer's name and address
       b. Manufacturer's type designation
       c. Manufacturer's identification reference
       d. Rated voltage
       e. Rated continuous current
       f. Interrupt rating
       g. Rated frequency
    6. In addition to the above manufacturer's nameplate, furnish a laminated plastic nameplate in accordance with the Identification Section.
    7. Panelboards shall be complete with bolt-type wire connectors.
2.03 Buses
  A. Buses in the panelboard shall be of copper sized in accordance with Underwriters Laboratories standards. Connections shall be bolted and laminations interleaved to secure maximum contact areas. All buses and stub connections shall be made of such a size as to limit the

temperature rise to 50°C when carrying full-load current capacity. Full-size insulated neutral bars shall be included. Bus bar taps for panels with single pole branches shall be arranged for sequence phasing of the branch circuit devices. Bussing shall be braced throughout to conform to industry standard practice governing short circuit stresses in panelboards, but in all cases, bracing shall be equivalent to the rated interrupting capacity of the smallest circuit breaker in that panelboard, or 10,000 A (or as shown on the drawings) rms symmetrical at maximum rated voltage for panelboards having no breakers. Phase bussing shall be full height without reduction. Cross connectors shall be copper.
   B. Neutral bussing shall have a suitable lug for each outgoing feeder requiring a neutral connection.
2.04 Boxes
   A. Boxes shall be made from code gauge steel having multiple knockouts unless otherwise noted. Surface-mounted boxes shall be painted to match the trim. Boxes shall be of sufficient size to provide a minimum gutter space of 4 in on all sides.
   B. At least four interior mounting studs shall be provided.
   C. Panels shall be not more than 60 in high.
2.05 Trim
   A. Hinged doors covering all circuit breakers or switch handles shall be included in all panel trims. In making switching device handles accessible, doors shall not uncover any live parts.
   B. Doors shall have flush-type cylinder lock and catch, except that doors over 48 in in height shall have a vault handle and three-point catch complete with lock, arranged to fasten door at top, bottom, and center. Door hinges shall be concealed. Two keys shall be supplied for each lock. All locks shall be keyed alike.
   C. The trims shall be fabricated from code gauge sheet steel.
   D. Trims for flush panels shall overlap the box by at least $3/4$ in all around. Surface trims shall have the same width and height as the box. Trims shall be fastened with quarter turn clamps.
2.06 Circuit Breaker-Type Panelboards
   A. Circuit breaker-type panelboards shall be equipped with circuit breakers with frame size and trip settings as shown on the Drawings.
   B. Circuit breakers shall be molded case, plug-in type

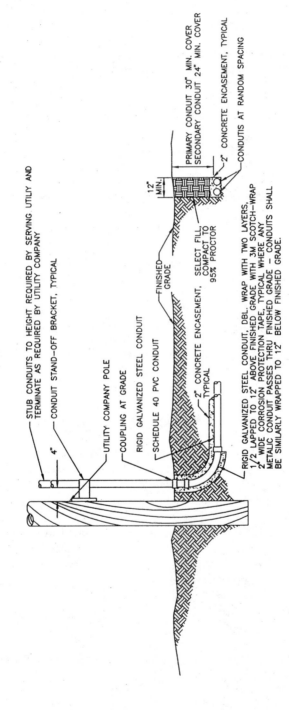

Figure 18.7 Primary conduit riser detail.

C. Breakers shall be thermal magnetic type employing quick-make and quick-break mechanisms for manual operation as well as automatic operation. Automatic tripping shall be indicated by the breaker handle assuming a distinctive position from manual "on" and "off." All multiple breakers shall have a common trip. Tie handles will not be permitted.
D. All circuit breakers 225-A frame and larger in power panels shall have interchangeable trips and all breakers shall have provisions for padlocking in the open position.
E. Circuit breakers used in 120/208-V panelboards shall have an interrupting capacity of not less than 10,000 A, rms symmetrical at 240 V.

Part 3: Execution
3.01 Installation
   A. Install box, trim, and interior rigid and plumb. Center interior with door opening.
   B. Install panelboards with the top trim 6 ft from the finished floor, unless noted otherwise in the drawings.
   C. Field check all panelboard loading and reconnect circuits as required to provide balanced phase and line loads.
   D. Cables installed in wiring gutters of panelboard shall be neatly bundled, routed, and supported. Minimum bending radius as recommended by the wire and cable manufacturer shall not be reduced.
   E. Check and verify torque connections on all panel bus connections in accordance with the manufacturer's recommendations.

VII. Lighting Equipment and Lamps Section

Part 1: General
1.01 Description
   The work included in this section of the specifications includes furnishing of all materials, labor, and equipment, except as furnished under other sections of the specifications or as noted on and specified herein. The work in general shall consist of, but is not necessarily limited to, the following:
   1. Lighting fixtures
   2. Ballasts
   3. Poles
   4. Lamps
1.02 Quality Assurance
   A. Reference standards
      1. National Electric Code, latest edition

2. NEMA Standards
3. Underwriters Laboratories
B. All material and installation methods used shall be in full accordance with the latest and approved electrical and mechanical engineering practices. All material furnished by the Contractor shall be new and bear inspection labels of the Underwriters Laboratories.
C. All work and materials shall be in compliance with the rules and requirements of the Underwriters Laboratories and the National Electric Code.

1.03 Submittals·

Submittals Requirements: The Contractor shall be required to submit shop drawings.
1. Manufacturer's data sheets shall be submitted as complete booklets, six copies required, for the following equipment:
   a. Fixtures
   b. Poles

Part 2: Products

2.01 Materials
A. Fixtures will be furnished as indicated on the drawings. No fixtures, other than those specified will be allowed.
1. All fixtures shall be furnished complete with all fittings, parts, and stems as required by the particular installations. All fittings, parts, and stems shall be of the same manufacturer as the fixture on which they are used and shall be installed strictly according to the manufacturer's recommendations and/or as specifically detailed on the drawings. Any deviations from these specifications without specific approval from the Owner will be remedied without charge by the Contractor. All fluorescent ballasts to be electronic, with maximum 10 percent distortion.
2. Furnish all mounting devices and miscellaneous connection devices necessary for complete equipment installation.
3. Where canopies are installed on sloping ceilings, provide ball aligner canopies.
B. Ballast and lamp match-up shall be compatible on high-pressure metal halide fixtures. The Contractor shall install daylight lamp types.

Part 3: Execution

3.01 Installation

A. This Contractor shall furnish and install a lighting fixture as hereinafter specified and scheduled, on each and every outlet in accordance with the type designation shown on the Drawings. If a type designation is omitted, the fixture shall be of the same type as is shown for rooms of similar usage and size. Verify before purchase and installation.
B. Fixtures shall be located in coordination with the aisle patterns and as approved by the Owner.
C. Canopies shall be in contact with ceiling, and stems shall be plumb. Fixtures shall be in vertical and horizontal alignment.
D. Fixtures shall be suspended from, and supported by, the building structural system, using hanger rods or galvanized wire, so that no weight of the fixture is borne by the ceiling or ceiling suspension system. Wood supports between joist for supporting fixtures will not be acceptable.
E. Immediately before final inspection, this Contractor shall thoroughly clean all fixtures, inside and out, including plastics and glassware, shall adjust all trim to properly fit adjacent surfaces, replace broken or damaged parts, and light and test all fixtures for electrical as well as mechanical operation.

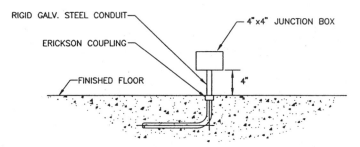

Figure 18.8 Exposed conduit stub-up.

Chapter 19

Electrical Checklist

A good checklist is the best way to have consistent and thorough plans. This list can be used by management for quality control of subordinate's work, by contractors to ensure all important items are included in the bid for services, and by engineering designers to check their own work. Some of the good things that will result from this level of quality control are

- Lower bid prices because there are fewer unknowns in the bid documents
- Faster and more trouble-free completion of work during construction
- Fewer bothersome change orders
- Quicker permitting of plans by the authorities
- A strong reputation for the department and the firm, for creating quality plans

Each individual, and office, has a different skill level and expertise. Quality control is constantly changing as staff changes, skills change, and the scope and size of projects change. The checklist should be used for every project and also updated with each project. If the list is used by a designer, the errors made in a submission should be added to the list. Department heads can add to their list those items that are tagged by the permitting authority, contractors in the bid process, or change orders that arise in the course of a project. The following is a good starting point for the list:

1. Outlets at 18 in above finished floor (AFF), per the Americans with Disabilities Act (ADA).
2. Switches, thermostats, and other control devices at 42 in AFF, per ADA.

3. Maximum of either duplex outlets per circuit, fewer if specific equipment ratings are known.
4. Key label all known equipment, including computers, copiers, printers, and other office or specialty equipment in addition to hvac, plumbing, and other mechanical items.
5. Emergency lights in corridors, with the symbol "E" added to the fixture type designation as described in a note to the fixture schedule.
6. All zonal cavity lighting calculations done on computer, using the latest floor plan, and with a computer printout of the results on file. The following criteria should be included in the analysis:
   - Actual wall, ceiling, and floor reflectances if known.
   - Increased ambient light level to account for partitions, dark furniture, paneling, bookshelves, and other such items that may significantly affect the lighting level.
   - Actual lamps to be used in the fixtures, with unique maintenance factors.
7. Exit signs marking all means of egress, with arrows shown where required and the proper symbol for ceiling or wall mounting as appropriate.
8. Photocell of exterior light fixtures.
9. Three- (or four-) way switches at each entrance to a space, as appropriate.
10. Locate switches controlling public lighting in staff areas to prevent inadvertent tampering.
11. One-line diagram with all wire sizes, disconnects, circuit breakers, and other power elements clearly labeled.
12. Calculate voltage drop (maximum 2 percent) on all home runs over 70 ft long, all feeders, and all wire sizes over #10 in all circumstances. Include the computer printout in the project design file.
13. Short circuit analysis on all services over 1000 A and for all services for hospitals and other facilities with public safety or health concerns at risk. In most cases a simple coordination analysis will suffice.
14. One-quarter-inch scale plan of electric and mechanical rooms with distribution power equipment shown with the major mechanical and plumbing equipment.
15. Any wire sizes other than the standard size (e.g., 3 #12, 1 #12G in ¾-in conduit) keyed clearly on the plans and the riser or one-line diagram as appropriate.

16. Junction boxes and pigtails to fixtures shown on the lighting plans, with switch legs shown dashed.
17. Complete MDP schedule with equipment ratings shown.
18. Equipment limitations on all major items, per specific quality manufacturers.
19. No disconnects mounted on equipment itself but on remote unistrut or kendorf galvanized racks.
20. Proper clearance in front of all equipment (e.g., 3 ft 0 in in front of panels), disconnecting means for all major hvac equipment in line of sight and within 50 ft 0 in. Such code requirements as these are always minimums. Four or five feet of clearance is better, and installing the disconnect switch within reaching distance of the equipment controlled is best. Use circuit breakers for the disconnecting means where possible, in small mechanical rooms.
21. Use a "main lugs only" type panel instead of a "main circuit breaker" panel if the panel is fed from a fused disconnecting means.
22. Specify mounting brackets or other nonstandard hardware for lights located in hard ceilings, on walls, or on poles.
23. On small jobs if specific light fixtures are not scheduled, include a reasonable lighting allowance.
24. On remodels always research the existing service and the facility use history to determine if the service can handle the additional load planned. If this is not done, clearly state on the plans that it is the contractor's responsibility to upgrade the service if necessary.
25. Do not ever allow conduits to be used as grounding means; always specify a ground wire for each circuit. Panels should also have a full-sized ground and neutral bus, properly grounded per code.
26. Provide full and complete symbol, lighting, and abbreviation schedules early in the set of plan documents for easy referral and reference.
27. Copy cut sheets for all lights, panels, and other major equipment in the project binder, keyed to the plan schedules used.
28. Coordinate equipment locations and designations with the mechanical and plumbing trades.
29. Provide power to specialty items such as VAV boxes, fire/smoke dampers, automatic dampers, heat tape, or heat tracing.
30. Show all smoke detectors, duct-mounted as well as ceiling-mounted devices, and all other safety devices requiring power.
31. Provide a minimum of one 4- × 8-ft plywood panel in the telephone/

communication closet, painted with two layers of nonconductive paint and have a grounding rod nearby.

32. Ensure that hvac, fire, controls, and other panels have dedicated circuits of sufficient capacity, for both present use and possible future expanded capacity of the panel or controls capability.
33. Include at least 10 percent spares in each panel and 20 percent spaces. The feeder should be increased in size to handle these additional loads.
34. Connect to all equipment, inside or outside, with a minimum of 4 ft 0 in of flexible, liquid-tight, metal conduit for vibration isolation and ease of maintenance.
35. Provide maintenance overrides on all VFDs so that the motor can continue to be used (at full speed) if the VFD fails or is removed for maintenance.
36. On all projects, new and remodel alike, show a thorough detail of the service entrance, to include Weatherhead or underground service feeders, CT can, meter, wire, and conduit sizes.
37. Provide a duplex outlet at each rooftop unit location.
38. Provide heat tracing circuit for all outside hvac water lines (e.g., to a cooling tower).

The next section looks at the checking of electrical work from another, more formal perspective. It is more appropriate for larger projects for less-trusted clients, although the method is applicable to small projects as well.

Division 16—Electrical Specifications

The written binder of specifications is supposed to augment the blueprints by providing additional detail on specific equipment and systems. The specifications should (1) establish a minimum quality for all electrical devices by identifying specific allowable manufacturers or by requiring a UL label or other standard certification, (2) describe installation procedures and standards, (3) ensure competitiveness in bidding by allowing several reputable manufacturers to bid the work, and (4) delegate to the contractor the requirement to provide and install complete systems even though the plans are only diagrammatic. The specifications are often quite general in nature and most sections can be copied and reused from job to job with little modification.

When unusual, complex, or costly equipment is to be installed on a project such as switchgear or an automatic transfer switch, the specification section should be very concise and detailed. Every feature, ca-

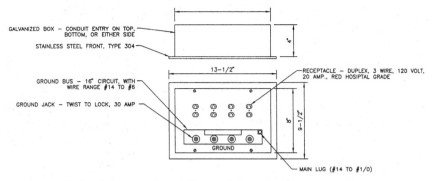

Figure 19.1 Isolated power and grounding module.

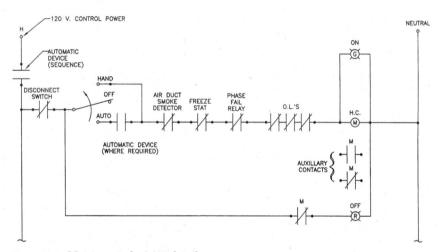

Figure 19.2 Motor control wiring detail.

pability, capacity, and rating should be identified exactly in the written specification. Most manufacturers include sample specifications in their catalogs, and these are a good starting point. Once these typical specifications have been created, it is a good idea to have another manufacturer's representative review them to ensure that they will be able to bid on the item.

One of the best ways to create a quality specification section is to purchase a canned master specification that can be easily modified to suit the job conditions. The best product on the market is a Windows-based package called Speclink (Building Systems Design, Inc.). The well-conceived program has many limits, checks, and cross-references built into the software to help the designer and to reduce the possibility of any conflicts in the specification and identification of electrical system components.

Following is a list of the most common specification sections by number. It is important to include specification sections for every unusual such item installed on a project. Also the list is a good review check for quality control, to be sure all of the important items on a project have been addressed either on the plans or in the specifications.

Number	Title
16060	*Grounding and Bonding.* Basic materials and methods for electrical systems, the building structure and piping, allowable grounding means for circuits and panels and other major equipment
16070	*Hangers and Supports.* Straps, clamps, channel, and other hardware to support electrical equipment along with spacing and weight limits
16075	*Electrical Identification.* Nameplates and labels for conduits, equipment, and other devices including type, size, color, and means of attachment
16095	*Minor Electrical Demolition.* Removal of wiring and conduit from demo areas and the proper safe termination thereof at junction boxes or panels
16121	*Medium-Voltage Cable.* Cable and accessories for systems rated 600 to 35,000 V
16123	*Building Wire and Cable.* Cable and accessories for systems rated up to 600 V
16125	*Undercarpet Cable.* Installation and use of flat conductor cable for power and signals circuits installed under carpeting or other flooring types
16127	*Manufactured Wiring Assemblies.* Ready-made wiring assemblies (conduit plus wiring plus ground) for use in home runs, pigtails for lights, and other branch circuits
16131	*Conduit.* Metal and nonmetallic conduit and tubing, fittings, and finishing criteria including plastic bushings on open-ended conduit
16132	*Surface Nonmetallic Raceways.* Surface-mounted PVC and other types of wireways as manufactured by Wiremold or Carlon
16132	*Surface Raceways.* Multioutlet power and signals wireways and wall-mounted electrical duct
16133	*Underfloor Ducts.* Installation guidelines and specification of fittings for the underfloor distribution of power and signals wiring
16134	*Cable Trays.* Metal and fiberglass cable tray design and installation, to include any special requirements for fiber optic installations as applicable
16135	*Utility Columns.* Indoor utility columns for power, telephone, and data/communication outlets including means of attachment

above and below and flexibility for movement within the ceiling grid

16138	*Boxes.* Pull and junction boxes, wall and ceiling fixture, and outlet boxes including accessibility from occupied spaces as necessary
16139	*Cabinets and Enclosures.* Metal cabinets for terminal blocks, main service entrance, and other electrical devices, including provisions for grounding, front and side access, finish, and locking capability
16140	*Wiring Devices.* Outlets, receptacles, wall switches, dimmers, telephone jacks, and access boxes in floors or other specialty areas
16155	*Equipment Wiring.* Connection for mechanical and architectural equipment such as modular furniture
16210	*Electrical Utility Services.* 600-V or lower service entrance details, equipment, wiring, and aid to construction fees
16215	*Electrical Sensing and Measurement.* Instrumentation meters, relays, switches, and transformers to monitor individual equipment or characteristics of each phase of the incoming service
16231	*Packaged Engine Generators.* Self-contained fuel oil or natural gas engine with a generator set and automatic transfer switch. Coordinate with mechanical for indoor installations that require combustion air and ventilation air
16243	*Emergency Power Supply.* Small battery or inverter power sources under 10 kVA
16261	*Converters.* AC-to-dc converters, frequency changers, and single-to-three-phase converters
16263	*Static Uninterruptible Power Supply.* Above 10 kVA but less than 300-kVA emergency power sources
16271	*Pad-Mounted Distribution Transformers.* Liquid-filled, pad-type transformers from 2 to 34.5 kV down to 600 V or less
16272	*Dry-Type Transformers.* Small enclosed transformers 600 V and below, for power and lighting loads
16281	*Power Factor Capacitors.* Capacitors for power factor correction, assemblies, and automatically switched, 600 V and below
16311	*Overhead Line Materials.* Conductors, insulators, and related accessories for outdoor overhead power distribution lines
16341	*Circuit Breaker Switchgear.* Switchgear rated 2 to 15 kV, with air-magnetic or vacuum interrupter circuit breakers
16342	*Air Interrupter Switches.* Air interrupter switches for applications of 34.5 kV or below
16343	*Medium-Voltage Oil Switches.* Oil switches for applications of 34.5 kV or below
16344	*Medium-Voltage Motor Controllers.* Fused, NEMA Class E2 motor controllers rated 6.6 kV or below
16360	*Unit Substations.* Secondary unit substations with primary voltage to 34.5 kV and low-voltage secondary

16411	*Enclosed Circuit Breakers.* Molded-case circuit breakers, current limiters, and trips
16412	*Enclosed Switches.* Disconnecting means (fused or not) for equipment, rated 600 V or below
16413	*Enclosed Transfer Switches.* Automatic and manual-type transfer switches, rated 600 V or below (integral to the emergency generator set is ideal)
16414	*Remote Control Switching Devices.* Low-voltage switching devices or panels for control of 120- or 277-V lighting or other applications
16422	*Peak Load Controllers.* Controllers to monitor usage and demand and to either signal alarm or automatically turn equipment off to limit the peak load
16423	*Enclosed Motor Controllers.* Manual or automatic controllers for ac induction motors
16424	*Motor Control Centers.* Enclosure for several motor controllers, starters, or VFDs for ac induction motors
16425	*Variable-Frequency Controllers.* Three-phase motor controllers of the pulse-width modulated design, including maximum allowable harmonic distortion and minimum device efficiency at several loading conditions
16426	*Enclosed Contactors.* For feeders and power or lighting branch loads
16442	*Distribution Switchboards.* Main service and distribution panels with enclosed circuit breakers and fused switches
16443	*Panelboards.* Load centers, panels with switches, and circuit breakers
16451	*Feeder and Plug-In Busway.* 600 V and below busway for 150- to 5000-A loads
16470	*Power Distribution Units.* Power conditioning equipment for sensitive electronic equipment
16491	*Fuses.* Fuse characteristics for devices to be used in low-voltage power circuits
16510	*Interior Luminaires.* Luminaire characteristics including emergency lighting circuits or battery packs, exit signs, and installation accessories
16520	*Exterior Luminaires.* Luminaires, poles, low-voltage controls, remote ballasts
16526	*Obstruction and Landing Lights.* FAA-specified marking, helopad landing and signal lights
16555	*Theatrical Lighting.* Fixtures, dimming and control equipment, and accessories
16721	*Telephone Service, Pathways, and Wiring.* Termination backboards and cabinets, wiring as applicable
16722	*Nurse Call System.* Call stations, enunciators, master stations, PA or intercom systems, power supply

Electrical Checklist 307

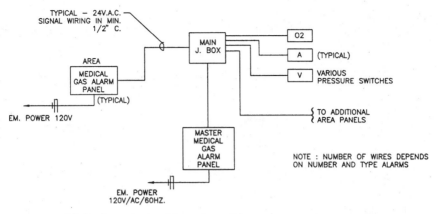

Figure 19.3 Medical gas alarm wiring schematic.

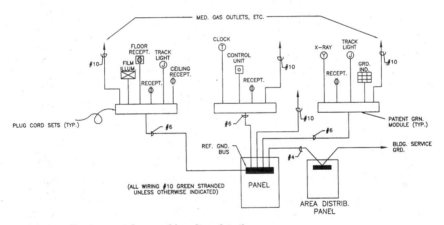

Figure 19.4 Equipotential ground bonding detail.

16723   *Intercom System.* Equipment, master stations, private interoffice capabilities

16781   *Television Distribution System.* Master antenna, satellite, and cable antenna

16821   *Public Address and Music Equipment.* Speakers, amplifiers, players, cables for paging, music, and announcements

Division 0 Sections/Checklist

Sometimes it is a good idea to clearly specify some aspects of the bidding process, especially for large or public projects. For example, for a large project that has changed constantly through the design process, it may be advisable to request unit prices from the contractor to cover

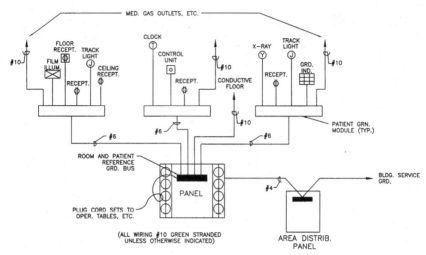

Figure 19.5 Equipotential ground bonding with isolation panels.

expected change orders. It is also acceptable to request additional information from bidders that will help in the evaluation of their capabilities, such as a list of equipment, current jobs, references, and other such information:

.00100	*Bid Solicitation.*	Invitation or advertisement to bid
.00200	*Instruction to Bidders.*	Bid security, acceptance criteria for bids
.00300	*Information Available to Bidders.*	Related documents availability and access, site visits, prebid walk-through
.00410	*Bid Form.*	For entry of bid amount, time, alternates, subcontractors
.00430	*Supplements to the Bid Form*	
.00431	*Supplement A—List of Subcontractors*	
.00432	*Supplement B—List of Unit Prices*	
.00433	*Supplement C—List of Bid Alternates*	
.00434	*Supplement D—List of Separate Prices*	
.00435	*Supplement E—Cost Breakdown*	
.00436	*Supplement F—List of Supplementary Information*	
.00437	*Supplement G—List of Equipment*	
.00438	*Supplement H—List of Tax Rebate Items*	
.00500	*Agreement*	
.00700	*General Conditions*	
.00800	*Supplementary Conditions*	

General Requirements

.01100 *Summary.* Contract scope and description, work not shown on drawings

.01200 *Price and Payment Procedures.* Applications for payment, closeout procedures

.01210 *Allowances.* Procedures and scope

.01230 *Alternatives.* Description and procedures

.01270 *Unit Prices.* Procedures and description of items to be included in the pricing

.01300 *Administrative Requirements.* Submittals, reports, meetings, schedules, coordination, availability of drawings on AutoCAD for creation of shop drawings

.01315 *Mechanical and Electrical Coordination.* Assigned to an individual, drawings requirements

.01400 *Quality Requirements.* Testing, inspections, reports, certificates, test reports, minimum UL requirements for all items not otherwise specified

.01425 *Reference Standards.* Full title and edition date for all national or local standards that apply

.01500 *Temporary Facilities and Controls.* Utilities, facilities, services for contractor's optional use

.01510 *Temporary Utilities.* Location, capacity, safety precautions

.01525 *Field Offices.* Location, phone, pager, or fax access

.01550 *Vehicular Access and Parking.* Quantity of parking places, location, penalties

.01565 *Security Measures.* Contractor procedures

.01585 *Project Signs.* Size, location, and description

.01600 *Product Requirements.* Options, substitutions, deliver, storage, protection, and bonding

.01700 *Execution Requirements.* Installation requirements, demolition, cutting and patching, cleaning, starting of systems, repairs, painting not elsewhere specified

.01780 *Closeout Submittals.* As-built drawings, record documents, operating and maintenance (O & M) manuals, warranties, bonds

Chapter 20

Designing for Conservation

This chapter offers some proven energy conservation guidelines for the design of lighting and other electrical systems. They work quite well for quality assurance for any project. They also can serve as incentive or rebate programs within a large organization that has many facilities. Finally, they provide some short-cut design methods that will not only get the design work done, but in such a way as to conserve energy to the best effect as well.

Lighting Power Limits

It is very easy to design a quality lighting system for a building, using stock lights, in reasonable quantities. The following design criteria is quite easy to satisfy, in almost every conceivable circumstance—including libraries, outside car lots at dealerships on "motor mile," and schools. Yet, the enforcement authority that generated this standard has had to become, by their own admission, very aggressive in making engineers comply with this really very lenient standard. For example, the most energy-intensive lighting design challenge of all, a library with high ceilings, can pass the standard with ease by a margin of 20 percent.

Evidently lighting designers are not doing even the most rudimentary lighting calculations, not even when they know very well the city will be reviewing their work. Perhaps it's a testament to the laziness of electrical designers and to their unwillingness to perform any calculations that these lighting power limits cause such consternation; or perhaps it's simply evidence of a paucity of good calculation programs available with which to perform the analyses.

For lack of a better guideline, the following method does limit the design to an energy-efficient one. It also, in the process, restricts the lighting selections to those within the range of the IES standards because the allowable watts limits the illumination levels to that criteria. It is a good office standard to follow to ensure energy-efficient designs and a uniform consistency to the design documents issued by the firm.

The Method

The first task is to fill out the data in a simple nine-column calculation form. The first three columns are for room function, ceiling height, and room area. The last three columns can also be filled out, with wattage and quantity of fixture and total watts in the room. The middle three columns of the form are extracted from tables.

The area factor is a function of the geometry of the space. This factor is extracted from Table 20.1 using the room area and ceiling height. The power density factor is a function of the space use. These values are dependent on the room use and are tabulated in Table 20.2. Multiplying the room area by these two functions generates the allowable watts in the space, which is tabulated in column 6.

The total of column 6 is the allowable lighting watts in the building. This is the maximum connected lighting load and must exceed the total of column 9. Individual rooms can exceed their limits, but the cumulative cannot surpass the total allowable budget. There are several general guidelines that must be followed to meet this standard:

- Use fluorescent lights in most spaces, with incandescent and spot lights used only in a few, special places for effect.
- Adhere to the IES illumination levels versus space use and occupancy.
- Use HID light sources in high ceiling spaces because fluorescent lights are too inefficient if the ceiling height is above about 14 ft.

These are common practices anyway, so the standard does not place any unusual demands on the designer.

Rebate Programs

This section discusses several rebate programs for electrical equipment, notably motors and lights. These programs can be used to evaluate new equipment purchases and energy-saving projects and the relative amount of money they will cost or save. Instead of spending a great deal of time performing a full energy analysis, the following programs will permit a quick and fairly accurate determination.

TABLE 20.1 Area Factors versus Ceiling Height (ft)

Area (ft^2)	8	8.5	9	10	11	12	14	16	18	20
50	2.00									
60	1.90	2.00								
70	1.80	1.92	2.00							
80	1.72	1.82	1.94							
90	1.66	1.73	1.85	2.00						
100	1.61	1.69	1.79	1.96						
110	1.56	1.64	1.73	1.91						
120	1.53	1.60	1.68	1.85	2.00					
130	1.50	1.57	1.64	1.80	1.97					
140	1.47	1.57	1.76	1.92	1.92					
150	1.44	1.51	1.57	1.72	1.87	2.00				
160	1.42	1.40	1.55	1.60	1.83	1.99				
170	1.40	1.46	1.52	1.65	1.79	1.94				
180	1.39	1.44	1.50	1.62	1.75	1.90				
190	1.37	1.42	1.48	1.60	1.72	1.86				
200	1.36	1.41	1.46	1.57	1.70	1.83				
220	1.33	1.38	1.43	1.53	1.65	1.77	2.00			
240	1.31	1.35	1.40	1.50	1.60	1.72	1.97			
260	1.29	1.33	1.38	1.47	1.57	1.67	1.91			
280	1.27	1.31	1.36	1.44	1.54	1.64	1.86			
300	1.26	1.30	1.34	1.42	1.51	1.60	1.81	2.00		
350	1.23	1.26	1.30	1.37	1.45	1.54	1.72	1.93		
400	1.20	1.23	1.27	1.34	1.41	1.40	1.65	1.83	2.00	
450	1.18	1.21	1.24	1.31	1.37	1.44	1.59	1.76	1.94	
500	1.17	1.19	1.22	1.28	1.34	1.41	1.55	1.70	1.87	2.00
550	1.15	1.18	1.21	1.26	1.32	1.33	1.51	1.65	1.65	1.99
600	1.14	1.16	1.19	1.24	1.30	1.35	1.48	1.61	1.75	1.91
700	1.12	1.14	1.17	1.21	1.26	1.31	1.42	1.54	1.67	1.81
800	1.10	1.13	1.15	1.19	1.24	1.28	1.38	1.49	1.60	1.73
900	1.09	1.11	1.13	1.17	1.21	1.26	1.35	1.45	1.55	1.66
1,000	1.08	1.10	1.12	1.16	1.19	1.24	1.32	1.41	1.51	1.61
1,500	1.05	1.06	1.07	1.10	1.13	1.17	1.23	1.30	1.37	1.45
2,000	1.02	1.04	1.03	1.07	1.10	1.13	1.18	1.24	1.30	1.36
2,500	1.02	1.02	1.03	1.05	1.00	1.10	1.15	1.20	1.25	1.30
3,000	1.00	1.01	1.02	1.04	1.06	1.08	1.12	1.17	1.21	1.26
4,000		1.00	1.00	1.02	1.04	1.05	1.09	1.13	1.17	1.20
5,000				1.01	1.02	1.04	1.07	1.10	1.13	1.17
6,000				1.00	1.01	1.02	1.05	1.08	1.11	1.14
7,000					1.00	1.01	1.04	1.07	1.09	1.12
8,000						1.01	1.03	1.06	1.08	1.11
9,000						1.00	1.02	1.05	1.07	1.19
10,000							1.02	1.04	1.06	1.08
20,000							1.00	1.00	1.01	1.03
30,000									1.00	1.00

TABLE 20.2 Lighting Power Density Limits

Interior task or area	Power density (W/ft^2)
Common Areas	
Boiler room	0.6
Conference room	1.0
Corridor	0.5
Dining room (fast service)	2.1
Dining room (leisure)	1.6
Electrical equipment room	0.5
Garage, parking	0.2
General assembly (auditorium)	0.8
Kitchen	1.3
Laboratories	2.4
Library room	1.7
Lobby, reception, waiting	0.8
Locker room & shower	0.5
Mail room	2.1
Material handling (bulk)	0.6
Mechanical equipment room	0.5
Stairs	0.5
Storerooms, warehouses	
Inactive	0.2
Active bulky	0.4
Active medium	0.5
Switchboard & control room	1.3
Toilet & washroom	0.6
Utility room, general	0.4
Office	
Accounting	2.4
Drafting	3.5
Filing (active)	1.5
Filing (inactive)	0.6
Graphic arts	2.3
Office machine operation	
Computer machinery	1.3
Duplicating machines	0.6
EDT I/O terminal (internally illuminated)	0.5
EDT I/O terminal (room illuminated)	1.3
Typing & reading	1.7
Residential	
Bath	4.30
Bedroom	1.40
Finished living space	2.20
Garage	0.50
Kitchen	4.00
Laundry	1.00
Unfinished living spaces	0.50

TABLE 20.2 Lighting Power Density Limits (*Continued*)

Interior task or area	Power density (W/ft²)
Commercial & Institutional	
Armories	
Drill	0.5
Exhibitions	0.6
Seating areas	0.4
Art galleries	1.2
Banks	
Lobby, general	1.7
Posting & keypunch	3.5
Tellers' stations	3.5
Bar (lounge)	0.8
Barber & beauty shops	2.9
Church & synagogues, main worship area	1.7
Club & lodge rooms	0.8
Courtrooms	0.7
Depots, air terminals & stations	
Baggage checkrooms	1.0
Concourse	0.6
Platforms	0.5
Ticket counter	1.7
Waiting & lounge area	0.6
Hospitals	
Autopsy	3.5
Central sterile supply	1.4
Corridor, special areas	
Nursing areas	0.6
Surgical & lab areas	0.9
Critical care areas	4.3
Cystoscopy room	4.3
Dental suite	3.0
EKG & specimen room	0.9
Emergency outpatient	4.3
Endoscopy room	3.3
Examination & treatment	2.1
Fracture room	2.1
Inhalation therapy units	0.9
Laboratories	3.5
Linen room	0.8
Lobby (entrance foyer)	2.1
Medical illustration room	6.2
Medical records room	3.1
Morgue	0.7
Nurseries	4.3
Nurse's station	1.1
Obstetric delivery suite	
Delivery room	8.5
Labor room	0.8
Postdelivery recovery	1.3
Occupational therapy	1.4
Patient's rooms	0.9

TABLE 20.2 Lighting Power Density Limits (*Continued*)

Interior task or area	Power density (W/ft^2)
Commercial & Institutional	
Pharmacy	3.4
Physical therapy	2.1
Postanaesthetic recovery	4.3
Pulmonary function lab	3.5
Radiological suite	
Preparation area	2.8
Special procedures	5.6
Treatment room	0.4
Solarium	0.6
Surgical induction & hold area	3.5
Surgical suite	
Operating room, general	8.5
Scrub & cleanup area	2.8
Sterilizing & instruments	0.8
Utility room, work area	1.4
Waiting area, general	0.6
Hotels	
Bathrooms	1.0
Bedrooms	0.4
Entrance foyer	0.8
Lobby, general	0.8
Laundries	
Fine hand ironing	2.1
General processing	1.0
Washing	0.6
Library	
Audio listening areas	0.6
Audiovisual areas	1.3
Book stacks (active)	0.7
Book stacks (inactive)	0.4
Book repair & binding	1.4
Card files	2.4
Cataloging	1.7
Microfilm areas	1.7
Reading areas	1.7
Municipal building, fire & police	
Fire engine room	0.6
Firemans' dormitory	1.4
Identification records	3.5
Jail cells	0.6
Recreation room	0.7
Nursing homes	
Administration & lobby	1.1
Chapel & quiet area	0.7
Nurses' station	1.1
Occupational therapy	1.0
Patient care unit	1.1
Pharmacy area, general	1.2
Physical therapy	1.4
Recreational area	1.1

TABLE 20.2 Lighting Power Density Limits (*Continued*)

Interior task or area	Power density (W/ft^2)
Commercial & Institutional	
Post Offices	
Lobby	0.6
Sorting, mailing, etc.	2.1
Schools	
Art	2.3
Classrooms	1.7
Dormitories	1.1
Drafting	2.4
Home economics	1.1
Laboratories	2.1
Lecture	1.7
Music	1.3
Sewing	3.0
Shops	2.1
Study halls or typing	1.7
Service stations, auto	0.6
Stores	
Alteration & fitting	4.3
Circulation	0.7
Merchandise	2.9
Sales transaction	1.4
Show windows	6.5
Stockrooms	0.6
Wrapping & packaging	1.0
Theatre & movie houses	0.8
Industrial	
Aircraft maintenance	
Docking & maintenance	2.2
Engine overhaul	3.1
Fabrication	1.6
Aircraft manufacturing	
Assembly, sub & final	3.6
General production	2.6
Inspection assembly	3.6
Inspection stock parts	7.1
Testing	
Extra-fine instruments, scales	7.1
General	1.5
Textile mills	
Drying & finishing	4.3
General production	1.5
Warping, weaving, grading	4.3
Tobacco products	
General production	0.9
Grading & sorting	8.5
Upholstering	3.8
Welding	1.4
Woodworking	1.6

TABLE 20.2 Lighting Power Density Limits (*Continued*)

Interior task or area	Power density (W/ft²)
Sports	
Seating area—all sports	0.4
Badminton	
Club	0.5
Recreational	0.4
Tournament	0.6
Basketball	
College & professional	1.1
College intramural, high school	0.6
Bowling	
Approach areas	0.4
Lanes	0.5
Boxing or wrestling (ring)	
Amateur	1.9
Championship or professional	3.8
Exhibitions, matches	1.1
General exercising & recreation	0.6
Handball	
Club	0.6
Recreational	0.5
Tournament	1.1
Hockey, ice	
Amateur	1.1
College or professional	2.1
Recreational	0.5
Skating rinks	0.4
Swimming	
Exhibition	1.0
Recreational	0.6
Tennis	
Professional (class I)	2.1
Club (class II)	1.4
Recreational (class III)	1.1
Tennis, table	
Club	0.6
Recreational	0.5
Tournament	1.1
Volleyball	0.5
Exterior Areas	
Driveways	
Private (2-lane width)	2.0 (W/linear ft)
Public (2-lane width)	3.0 (W/linear ft)
Entrances, without canopy	30.0 (W/linear ft)
Loading doors	20.0 (W/linear ft)
Exits, with or without canopy	20.0 (W/linear ft)
Entrances with canopy	
Decorative (hotel, theatre, etc.)	10.0 (W/ft²)
Utilitarian (hospital, office, etc.)	4.0 (W/ft²)
Loading areas	0.3 (W/ft²)

TABLE 20.2 Lighting Power Density Limits (*Continued*)

Interior task or area	Power density (W/ft²)
Exterior Areas	
Outdoor production & processing	0.4 (W/ft²)
Outdoor storage	0.2 (W/ft²)
Parking lots	
Open, public	20.0 (W/space)
Open, private	30.0 (W/space)

SOURCE: Reprinted with permission from ASHRACE/IES from Standard 90.1.

Motors

Electric motors account for a major percentage of the electric power used at commercial and industrial facilities. Large motors are used in elevators, air handlers, chillers, cooling towers, exhaust fans, pumps, and many other instances. This analysis applies to the evaluation of either new or retrofit motors. It is highly recommended that a high-efficiency motor be purchased in all instances. The additional cost over a standard motor will pay for itself in only a year of use at 50 hours a week. The following are qualifications to apply for this rebate:

1. Purchase must be made of ac motors specified in Table 20.3 and of NEMA design A or B.

TABLE 20.3 Nominal Motor Efficiencies

		Open drip-proof			Totally enclosed		
Hp	Rebate ($)	1200 rpm	1800 rpm	3600 rpm	1200 rpm	1800 rpm	3600 rpm
1	10	77.0	82.5	-	75.5	80.0	—
1.5	10	82.5	82.5	80.0	82.5	81.5	82.5
2	10	84.0	82.5	82.5	82.5	82.5	82.5
3	10	85.5	86.5	82.5	84.0	84.0	84.0
5	15	86.5	86.5	85.5	85.5	85.5	85.5
7.5	25	88.5	88.5	85.5	87.5	87.5	87.5
10	30	90.2	88.5	87.5	87.5	87.5	87.5
15	50	89.5	90.2	89.5	89.5	88.5	89.5
20	70	90.2	91.0	90.2	89.5	90.2	89.5
25	90	91.0	91.7	91.0	90.2	91.0	90.2
30	100	91.7	91.7	91.0	91.0	91.0	91.0
40	150	91.7	92.4	91.7	91.7	91.7	91.7
50	200	91.7	92.4	91.7	91.7	92.4	91.7
60	250	92.4	93.0	93.0	91.7	93.0	91.7
75	300	93.0	93.6	93.0	93.0	93.0	93.0
100	350	93.6	93.6	93.0	93.0	93.6	93.0
125	400	93.6	93.6	93.0	93.0	93.6	93.0
150	500	93.6	94.1	93.6	94.1	94.1	94.1
200	600	94.1	94.1	93.6	94.1	94.5	94.1

2. The full-load efficiency of the motors purchased must be greater than those listed in Table 20.3. This efficiency must be rated according to NEMA guidelines.
3. The motors are required to have a minimum of 1-year warranty for parts and labor.

The rebate is calculated by the following conservative method. Generally, the expected savings will be at least 2 to 3 times this value. This method awards $1 for each percentage point of efficiency gain above the nominal standard efficiency. The function used is

$$\text{Rebate \$} = (\text{new efficiency} - \text{Table 20.3 efficiency}) \times \$30.00 \times \text{no. of motors} \quad (20.1)$$

Large institutions with many buildings can use this rebate method to implement a companywide energy conservation program, confident that the conservative results of this analysis will provide a very good return on the investment.

Lighting

Lighting consumes 30 to 40 percent of the energy used at a typical commercial building. Installation of more efficient lighting can reduce this by up to 50 percent. The retrofit will not substantially affect the quality of the lighting or the visual environment. The following qualifications must be fulfilled for each instance:

1. The lamps, ballasts, and fixtures must be disposed of properly and not reused. Ballasts that contain PCBs must be disposed of as a hazardous material, according to prevailing local, state, and federal regulations.
2. The scope of work must be clearly defined, and calculations performed (on the basis of ratio of total lumens of light output) on the basis of readings taken by a properly calibrated light meter.
3. All lighting must be shown to comply with the national lighting standards promulgated by the IES.
4. Exterior lighting must be controlled by a photocell.

Table 20.4 shows the performance of typical fluorescent fixtures.

Current limiters

The power consumption of existing fluorescent fixtures can be reduced by one-fourth to one-half by the installation of current limiters. The light output is reduced by about the same proportion as the re-

TABLE 20.4 Typical Fluorescent Fixture Performance

Description	Typical fixture	Wattage	Annual operating cost ($)
Fixture with standard ballast			
One 4-ft lamp	F40-40W	57	12.45
	F40-34W	50	10.92
Two 4-ft lamps	F40-40W	96	20.96
	F40-34W	84	18.34
Two 8-ft lamps	F96-75W	173	37.78
	F96-60W	138	30.13
With energy-efficient ballast			
One 4-ft lamp	F40-40W	50	10.92
	F40-34W	43	9.39
Two 4-ft lamps	F40-40W	86	18.78
	F40-34W	72	15.72
Two 8-ft lamps	F96-75W	158	34.50
	F96-60W	123	26.36
With high-performance ballast			
Two 4-ft lamps	F40-40W	69	15.07
	F40-34W	57	12.45
Two 8-ft lamps	F96-75W	130	28.39
	F96-60W	105	22.93
Three 4-ft lamps	F40-40W	105	22.93
	F40-34W	90	19.65
Four 4-ft lamps	F40-40W	126	27.51
	F40-34W	107	23.36

duction in wattage. It is best to use devices with a current crest factor of 1.7 or less and a power factor correction.

One of the qualifications for the installation is that the current limiter must be directly wired into the secondary side of the ballast. In addition, the power consumption of the lamp/ballast combination must be reduced by at least 30 percent as verified by a qualified, independent testing laboratory. A minimum 5-year warranty is standard.

The rebate is on a per-unit basis. Typically it is at the level of $2.50 per unit for standard devices and $4.50 per unit for high-performance ones.

Other projects

There are many other lighting projects that can be implemented at a typical facility. These usually obtain rebates on the basis of kilowatt reduction, typically at the rate of $100 per kilowatt.

Optical reflectors are shaped metal inserts to fluorescent fixtures that have a polished surface and optimum shape to alter the photometrics of the fixture. One or two lamps can be removed from the fixture after the installation of an insert. Typical installations are in common areas that presently receive excess light.

Often it is more economical to replace older fixtures completely. By the time the entire device is upgraded, including new lens, lamps, and ballast, it is cheaper to purchase packaged fixtures, shrink-wrapped with lamps and ballast already installed. The useful lifetime will be far longer and the efficiency superior.

Exit signs are a good retrofit application because they operate 24 hours a day. Incandescent lamps can either be replaced by fluorescent bulbs or by LED kits.

Chapter 21

The Production Process

The design of electrical systems for a building requires far more coordination with the client than any other engineering discipline. With the exception of the structural engineer, it also requires more coordination with the project architect than any of the other engineering disciplines. The electrical engineer's lighting design also has more impact upon the visual environment than any discipline except architectural, and, in some instances, more than even architectural. Finally, the impact of the lighting design upon the health and comfort of the occupants is a close second to the hvac design.

These are some compelling reasons for diligence on the part of the electrical designer. Even more importantly, they highlight clearly the need for a complete and insightful understanding of the client. The use of each space in the facility must be well known, even to the point of what piece of equipment is to be plugged into specific outlets, throughout the building. The seating of occupants in each space must be known, too, for proper orientation of overhead lights—and for placement of outlets for power, communications, and telephone use. The electrical engineer must even have an understanding of the future of the industry and how this will affect the client and the use of the space.

There are two common resolutions to this dilemma. First, the client interactions are delegated to the architectural team leader. Consequently all of the important information is obtained second hand, if at all. Second, instead of making an honest attempt to comprehend the needs of the client, the entire system is simply overdesigned. It's better to have a system that is overqualified than one that falls short. An active participation will avoid both of these pitfalls and result in a better-designed electrical system and a more satisfied client.

The Schematic Phase

This is the nascent phase of the project. The architect visits long and often with the client to develop a vision of the building, the space needs, and the use of the facility by the occupants. The engineering disciplines are seldom contacted, except to inquire about the location of equipment rooms and the availability of services on site.

The usual outcome of this process is that, along with the space use, all of the budget concerns are resolved as well. Because the project architect does all of the negotiations, all decisions concerning cost are made in favor of that discipline. Little matter that lighting, especially, can affect the design almost as much as architectural features—sometimes more so, in the case of simple, basic designs. Moreover, the clever use of lighting can achieve certain effects more economically than architectural nuances. It is the electrical engineer's charge to make such options known to the architect and to the client, and that means an active involvement in the project from the very beginning.

There are other aesthetic considerations, with respect to lighting, that affect the architectural footprint and elevations. The architect may desire large windows in the exterior offices or even an entirely glass exterior wall. This will give a dynamic feel from the outside and plenty of light inside. It is a common misconception that you can never get enough daylighting in a space. More practically, the large glass expanses will be provided with blinds, and the occupants will keep these blinds partially closed all the time. Also, the IES promulgates lighting levels for a purpose: too much light can be as distracting, even as harmful, as too little light. This should be a factor in the design of any spaces with exterior glass exposure not just to optimize the size of the windows but also to provide lighting controls to save energy and to keep the ambient light level in offices from becoming excessive when daylighting is available.

Other important questions need to be posed by the electrical designer of the client concerning general space use:

- Where computers are used and if they are networked
- Where other office equipment, such as fax machines, copiers, and other devices are used, and if they can be linked to the office network
- Use of kitchen appliances, in a break room or otherwise
- Special equipment items—specific cut sheets are helpful, or nameplate data on existing equipment to be relocated
- Use of modular furniture or open office arrangements, where power needs to be delivered via power poles, floor outlets, or hard wired at junction boxes

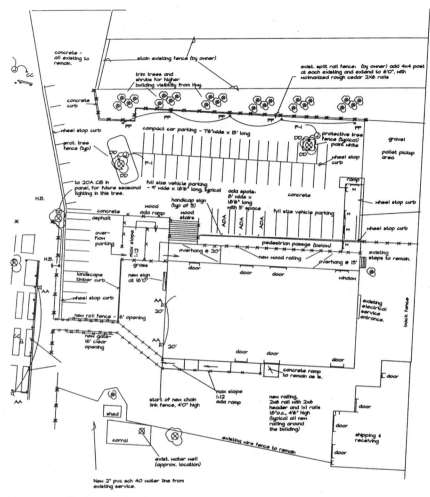

Figure 21.1 Garden center plan—site.

This is the sort of information that must be sought in face-to-face meetings with the client. If you ask the architect to collect it for you, the results are likely to be lackluster and incomplete. If this is not possible, a less direct approach is called for.

Most project architects have a schematic submission to the client, detailing the floor plan. The client is then asked to sign off on the drawings, confirming the layout for the continuation of work. This floor plan becomes the basis for all the subsequent engineering and architectural work, and any deviations from it are charged to the client as extra services. Thus it is a necessary step in any sizable proj-

326 Chapter Twenty-One

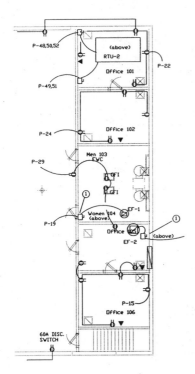

NOTES:

1. ALL TELE/COMM OUTLETS HAVE A PHONE JACK AND A COAX JACK PROVIDE 3/4" CONDUIT STUBBED OUT TO CEILING SPACE. FROM EACH J-BOX, WITH PULL STRING AND PLASTIC BUSHING.
2. ALL OUTLETS AT 18" AFF, UNLESS OTHERWISE NOTED.
3. PROVIDE MOTOR RATED DISCONNECT SWITCH FOR ALL AIR HANDLERS, EXHAUST AND VENT FANS IN THE ATTIC SPACE. INTEGRAL TO THE UNIT OR REMOTE, AS SHOWN HERE.
4. STUB UP RIGID CONDUIT IN WALLS FOR TELE/COMM.
5. ALL HOME RUNS NOT LABELED ARE TO BE ROUTED BACK TO THE ELECTRIC ROOM. PROVIDE A 20A CB FOR EACH.
6. PROVIDE A TWO SECTION PANEL 'P' WITH 20 30 CKT PANELS, 400 A MAIN CB.

```
P1 - 3    20A/1P    EXTERIOR LIGHTS
P4-12     20A/1P    POWER POLES AND OUTLETS
P12-17    20A/1P    SPACE FANS (NEW & EXIST)
P18-21    20A/1P    EXHAUST FANS 1,2,3
P22-30    20A/1P    DUPLEX OUTLETS
P31,33,35 70A/3P    RTU-1
P32,34,36 50A/3P    HTR-1
P-37,39   30A/2P    ELECRIC HEAT IN RTU-1
P-40-46   20A/1P    LIGHTS
P-47,49   30A/2P    ELECTRIC HEATER
P-48,50,52 50A/3P   RTU-2
P-49,51   30A/2P    ELECTRIC HEAT IN RTU-2
P-53-55   20A/2P    EF-2
P-54,56   30A/2P    ELECTRIC HEATER
```

7. MINIMUM CIRCUIT WIRING IS #12 WIRING IN 3/4" CONDUIT.
8. INSTALL POWER POLES WITH PLUG-IN CONNECTION ABOVE THE CEILING SO THEY CAN BE REMOVED EASILY. PROVIDE FOUR DUPLEX OUTLETS AND TWO JACKS (ONE FOR TELEPHONE, ONE FOR COMPUTER) ON EACH OF THE POWER POLES SHOWN.
9. SIZE ALL HOME RUNS, DISCONNECTS AND OTHER ELECTRICAL ITEMS PER THE NATIONAL ELECTRIC SHOWN.
10. LOCATE ELECTRIC HEATERS IN CORNERS OF THE MERCHANDISING SPACE, 4'0" BELOW THE CEILING.
11. MOUNT UNIT DISCONNECT FOR THE ROOFTOP UNITS AND THE HEATING ELEMENT ON A UNISTRUT RACK, WITH A GFI/WP DUPLEX OUTLET ALSO ON THE RACK.
12. PROVIDE TWO DUPLEX OUTLETS IN THE SECOND FLOOR SPACE ON A SINGLE 20A CIRCUIT.
13. REMOVE ALL EXISTING PANELS, CONDUIT, WIREWAYS ON THE INSIDE OF THE BUILDING AND THOSE ON THE OUTSIDE AS WELL SO LONG AS SERVICE TO THE BUILDINGS ON SITE TO REMAIN IS NOT AFFECTED.

KEYED NOTES

1. MOTOR RATED DISCONNECT SWITCH
2. 100A,NEMA3R DISCONNECT SWITCH, FUSED TO 70A ON

① POWER AND SIGNALS PLAN
1/8" = 1'0"

Figure 21.2 Garden center plan—power.

ect and a good opportunity for the electrical designer to get some feedback from the client.

Even the most basic schematic floor plan will have rooms labeled or use noted in some way. The electrical designer can on this basis alone locate duplex outlets and telephone/computer jacks in the most logical fashion. Even if it is just done by hand with a simple symbol key, it gives the client something to look at. These devices may not even

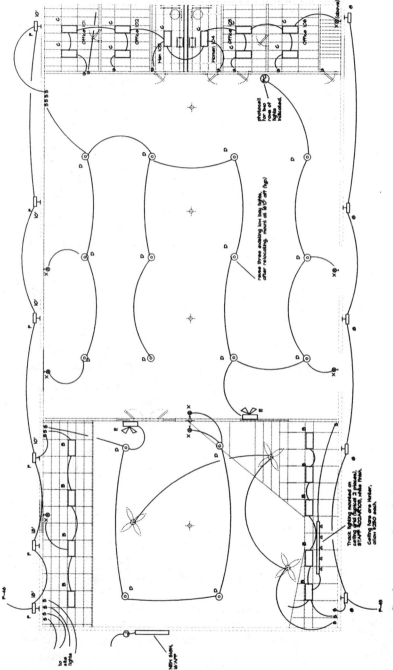

Figure 21.3 Garden center plan—lighting.

have come under consideration yet, and having even a preliminary layout is very helpful in visualizing the use of each space.

It is a good philosophy to provide the client with as much and as detailed information as you can as soon as possible in the project. Even at the risk of erring greatly and suffering embarrassment (or antagonism from the architect), the less refinement that is left until later in the project, the better. The schematic phase, architecturally, is a time for conceptualizing and visualizing. Later will be the time to design, cost estimate, and draft—the less conceptual design that is done at that point, the better.

A final, but very important point to discuss: budgetary constraints. A clear direction must be provided on the scope of the electrical work to be done. Paramount among these considerations is the budget for lighting. This is always the first item to discuss in value engineering a job back into budget and is not a good thing to change after the entire job has been designed. Changing fixture types has ramifications throughout the entire design: calculations must be redone, fixture quantities changed, circuits designed again, reflected ceiling plan reconfigured, hvac loads redone, and so forth.

Of course, this extent of redesign is never done, even if it should be. The result is a project that is very loose in design in all respects. The electrical design is the worse, and it remains so because there is rarely time or inclination to do it all over again. All of this can be avoided if the client is asked to sign off on a specific lighting budget, with the support and acknowledgement of the project architect if possible because of its strong impact upon the entire space. This simple contract will shift lighting modifications during the value engineering phase from first priority to last. It will also guarantee extra services if the fixtures are changed (e.g., from parabolic to fluorescent), so the electrical distribution system can be properly designed to completely fit the needs of the building, its equipment, and its occupants.

Design Development

The design development of a project can be considered to be all those activities that occur before the client approves the final floor plan. *Activities* is the key word here because most electrical designers take a very low profile role in this process. It is important that a much more assertive role be played. There are many reasons for this.

Increasingly, the onus to conserve energy is changing from the mechanical designer to the electrical engineer. The hvac design and equipment in many cases is so constrained by code requirements, which strictly govern energy use and outside air requirements, that there is little more the mechanical engineer can do than to just comply with the code. The same cannot be said of electrical equipment. The codes require a reasonable efficiency in light bulbs and motors, but little else. There remain many, many areas of electrical design open to energy-saving design strategies—lighting layout, lighting controls, circuit and home run design, and motor selection to name but a few. If a new building is to achieve an ambitious reduction in energy consumption, at least 50 percent will have to come from the electrical power distribution system.

This is not an easy task because very little of the equipment is actually subject to an independent selection by the electrical engineer. The lighting must be negotiated with the architect, the hvac fan and air handler motors with the mechanical engineer, and the hot water distribution and heating source with the plumbing engineer. The nonassertive solution is to let them choose what they will and then just install capacitors as required to lower the power factor to reduce energy consumption. This is also a tactic to use when all the other team members are uncooperative; the electrical designer is charged with reducing their energy use, in spite of their opposition.

Other than this gerrymandering of authority in equipment selection on the project, there are some concrete tasks required of the electrical designer in this phase of the project. It is best to provide the owner with a set of plans showing all of the design development (DD) work, even if it is not a part of the contract or if none of the other disciplines do the same. There is just too much important input that the client can provide to make it a much more satisfactory final project. At a minimum, the following items should be shown on the set of DD plans:

- Location of the service entrance and the main building switchgear
- Size and location of the electrical and communications rooms
- Location of all lights by general type (e.g., 2×2, 2×4, downlight) situated in the reflected ceiling plan, if at all possible
- Location of all room light switches
- Location of all power receptacles along with any specialty outlets already identified by the owner
- Location of all telephone and communication outlets

A good rule to follow with power and communication outlets is to provide one in every possible location. It is easier for an owner to delete them than to visualize the floor plan, then add them where they might be required.

When the plans are complete it is time to arrange a meeting with the owner and any department heads or other management personnel who have a solid knowledge of the work space requirements. Everyone should review the plans and note any changes needed to all systems. Of particular importance are any equipment items, no matter how small or inconsequential. Any special task areas should also be identified so that proper lighting, or switching, can be provided.

The DD plan meeting is also the proper time to introduce some important ideas to the client, such as dual switching in areas with daylighting, special fixtures, and even the potential future need for fiber optic cabling in the building. It is important, though, to get the main concepts of power and basic communications on paper first. Doing it all at once is a little too complex and confusing and will almost always result in a lower level of control complexity than is economical, or proper, for the installation. It is a key role of the electrical designer to case the owner into such decisions, educating him or her in the process, to see the wisdom of certain selections.

Another key decision the electrical engineer can actualize, as necessary, is the power source of large hvac equipment. Most hvac designers are lazy. They would much prefer to use electric heat in VAV boxes, for example, than to design a complex hydronic system. A DX system with electric heat is also much easier to design and install than one with a gas furnace. In either case the difference in energy use is astronomical, so much so that many progressive cities prohibit all electric heating means. A similar line of reasoning applies to packaged DX rooftop units versus a central plant with a chiller. The former is easy to design but much less efficient and usually more costly, too. Chillers can even be a point of contention: gas engine-driven chillers are cost competitive, widely used, and a great deal less expensive to operate where diesel or natural gas is available.

Taking the lead on such issues may seem like meddling, and it will certainly not be welcomed by the mechanical engineer. However, the electrical engineer must realize that when heating tasks are delegated to electrical sources, the rest of the electrical system suffers the consequences. The quality of light fixtures must be lessened, reducing the overall quality of the whole facility. Also the number of duplex outlets decreases and the number per circuit increases, reducing the future capacity and flexibility of the building. Thus, it behooves the electrical designer to ensure that alternate heat sources are investigated and carefully weighed in the decision making process.

Quality control

Every effort should be made to be sure the final bid set of documents is accurate, error free, and without any conflicts between the disciplines. Every single mistake will result in one of two things: a correction in an addendum to the bid documents or a change order during the construction process. The first is embarrassing, time consuming, and risky. The second is all of these plus costly and damaging to the reputation of the firm. The goal of every engineering design firm should be no addenda and no change orders, and those who produce accordingly should be recognized, awarded, promoted, and emulated.

The first objective of the engineering department should be to establish a regular hierarchy. The head of the department should have a thorough document review a week or two before the bid documents are printed—the final of many progress reviews and the most thorough. There should be few changes, given prior oversight, and there should be time to implement the changes prior to the printing of final documents.

Both the designer and the reviewer should have used a checklist to ensure that all the important things are done already. There are several checklists in Chap. 19 that can be used successfully, although each individual will need to develop one to suit the project types and personal aptitudes.

It is important that a second person other than the designer review the project documents thoroughly. If this is not possible, the designer should schedule a review after at least a few days completely off the project so that he or she will see the work from a new and fresh perspective. Another possibility is to ask other trusted designers or even a contractor to review the documents for a small fee. Just be sure the confidentiality of the client's documents is maintained and that the contractor does not get an extra advantage over the competition as a result of the early inspection.

One of the biggest problems in coordination is that often the mechanical equipment is not specified until late in the design phase, or additional items such as exhaust fans or ventilators are added at the last hour. The best way to avoid this is to at least locate the equipment, such as a condensing unit, and draw the disconnect and wiring to a specific panel or motor control center. Then when the actual size and capacity is known, all that is required is to add the wire, disconnect, and circuit breaker sizes. At worse a single sentence in an addenda will provide the sizes, rather than a complete plan or elevation and modifications to the riser.

Another efficiency measure is to prioritize the items done in the final crisis stage of the project, just before the drawings are issued. If a light, air handler, or fan is not in the exact location, that is less of a problem than if a disconnect is improperly sized or left out. The more costly the equipment or the more impact it has on life safety or important code issues, the higher the priority to correct it. It is best, in the final check, to start at the service entrance and proceed from there to the main distribution panel, the feeders, and eventually to the home run designations. Unless special equipment is used on the job, such as switchgear or a specific security or fire alarm system required by the owner, the specifications should have the last priority. This is especially true if the specifications are "recycled" from another, similar job and few things were changed.

Another strong incentive to pay special attention to the major items on the project is that the code review and permitting authorities typically focus mostly on the service entrance, riser diagram, and the load analysis. If they are all accurate and easy to follow, less attention will be paid to the balance of the plan work, and the permit will not be delayed. Also, contractors will not usually complain if a home run is mislabeled, a fan is out of place, or even a outlet is not circuited. They will, though, ask for compensation if a large disconnect is not shown or a feeder is undersized on the plans.

Bidding

The few weeks between the issuance of drawings and the submittal of bids from contractors are very important to the proper management of the project. Usually there are several calls by contractors with three different types of questions: clarifications, corrections, and revisions. In no case should additional information be given to any contractor that in any way alters the scope of the project. If such information must be issued, it should be done via an addenda item so that the exact same information is made available to all, at the same time.

There is always the temptation to play favorites with well-known contractors and to grant them subtle favors off the record. That is okay if it is only a matter of interpretation, but otherwise it is never a good practice. There are too many things that can happen in the interim to such informal agreements. The engineer of record may change, voiding the agreement. The contractor may misinterpret an agreement to consider an idea to be an acknowledgment thereof and underbid the project cost. There will be bad feelings when the truth must be implemented, if not worse, such as legal proceedings.

The best rule is just to never give off-the-record comments to anyone, friendly contractor or otherwise. It will not be held against you, in any event. Also it is a very good idea to keep conversation records of any phone conversations with contractors, equipment suppliers, or anyone else involved in the bid process. This will (1) establish a legal record that in fact no off-the-record agreements were made and (2) keep an accurate ledger of questionable items that need to be addressed in the addenda.

Another important part of the bidding process, especially for large projects, is the evaluation of potential bidders. Owners such as school districts that do a lot of construction work often have a prequalification procedure for contractors. This includes submittal of work history, current assets, references, and other pertinent information. Once a contractor is approved, other than occasional updates to the record, the firm is able to bid without further ado on all of the organization's projects.

Other, smaller organizations should also have a qualification process as part of the bid process so that a minimum quality of contractor can be assured. Legally such a process allows the owner to take bids other than the lowest one if it can be established that concerns for quality of work are warranted. On the down side, if the restrictions are too strict, the bidding may be less competitive and the bids may be higher than they should be. There is always a fine line between these extremes, and a savvy consultant is helpful in gauging the level of activity in the community and the potential pool of bidders.

Sometimes the process becomes so convoluted and the changes so drastic that the entire set of bid documents is discarded and the whole job is rebid. This is the worst possible outcome because it wrecks the owner's budget and timetable, aggravates the many contractors who have devoted considerable time taking off and estimating the project, and costs the project engineer great sums for the redesign. At a minimum, when any such debacle occurs on a project, the design firm should implement strict and enforceable measures to ensure that it never happens again.

Addenda

The most important thing to keep in mind about addenda is that they should never occur. When they do, it should be a clear sign to upper-level management that something is very, very wrong with the production process. If contractors in their brief and hasty review of the documents are calling in with more questions about the documents than can be answered off the record, the design engineer is not doing his or her job.

For those inevitable details that need formal correction in an addenda, it is important to identify the corrections clearly. If only a verbal description is used, a drawing number or specification section number should be described, plus careful and exact wording of the change promulgated.

Major changes may require a drawing on 8.5- by 11-in paper or sometimes an entire E or D size drawing. When drawings are reissued, the changes should be highlighted with a cloud and identified by the addenda number in a triangle symbol. In addition, for reissued full-size drawings, there should be a revision box in the title block for identifying the date, designer, and other pertinent information.

In every case the revisions are being made to the official engineered drawings that have been sealed by a professional engineer. The modifications must be made under his or her supervision and released only after proper authentication. Often the first page of an addenda is sealed, and when entire sheets are reissued, they should be resealed and dated by the original engineer of record.

Value Engineering

Again, value engineering is something that should never have to be done. If the designer was properly advised of the owner's financial limitations during the development phases of the project, and if the preliminary construction estimates were done correctly, there should be no need for further reductions in scope or cost, at least by the electrical design staff. Designwise, the only possible cost-saving measure should be in the type and quality of lighting fixtures; circuiting should not change as a result. All other aspects of the electrical design should either be done using the standard, cost-effective design approach provided to any client or be constrained by other factors beyond the control of the electrical designer.

Construction Management

A good set of drawings is the best way to ensure a trouble-free construction phase. With the exception of major equipment specifications, such as for switchgear or other exotic devices, the contractor should be confident to manage the entire installation from the blueprints. Field personnel are illiterate when it comes to reading specifications anyway, so if anything unusual that is not standard code behavior is required on the project, it should be clearly delineated on the drawings.

Rule number 2 in construction management is Listen to your contractor. If he or she recommends something, you can be assured that it is a good, economical idea. Unless it is an extremely disreputable contractor—and if the bid evaluation process was done right and references were contacted, this should never be the case—you can trust his or her judgment in almost all cases. The contractor has a reputation to keep up, no less than the engineer, and there is strong incentive to leave the owner with a favorable impression in all matters. The contractor's field personnel are seasoned professionals too and would not jeopardize their license by doing something against the code.

The next rule is to not be a stranger and try to manage a project from the office. Frequent, informal visits are a good way to not only keep apace of the project but to manage problems before they turn into change orders. Contractors have a well-developed aversion for paperwork and will do a little extra to avoid it. The engineer would do well to cultivate the same attitude.

The visits are particularly important early in a project and especially if it is a contractor you have not worked with before. A couple of walk-throughs and the contractor will become familiar with your standards, the higher the better, because everyone always relishes an opportunity to do quality work. An engineer who exhibits leadership and pride to the field people is solid gold to the contractor because it is an opportunity to train new, impressionable field hands. Do not shy away from the need to be a responsible and authoritative figure in the field; even if you struggle and flounder, your effort and perseverance will more than make up for inexperience.

Change Orders

One of the most helpful things that results from a good working relationship between engineer and contractor is in the area of the processing of change orders. There are several possible arrangements, depending on the relationship.

The most informal system applies for contractors who are well-known and trusted friends. If they request a change, you can trust it is needed. If small changes to the design documents are required to make the project work, usually he or she will make them without asking and do so at no extra cost to the owner. By the same token, they expect an honest deal from the owner and engineer or project manager to cover the cost for any major changes to the documents.

On small projects with such firms, it is possible to "save up" change orders through the course of the project. Informal agreements are made over the phone or with a hand shake, and the cost accounting is expected by both parties to be handled appropriately when the time comes. Often the changes balance over the project as a whole and no money or paperwork must change hands. A no-change-order project is like a baseball pitcher's no-hitter. It's a great way to build a reputation, for both the contractor and the engineer. In any event, it is still important to keep careful notes of all changes, agreements, and related information. This will keep both parties on an even keel in the event the field management changes during the course of a project and a full reckoning is called for.

The other extreme is government work. Usually every single item that is changed in any way from the construction documents must be itemized exactly, complete with drawings, cost documentation, and written approval from everybody from the foreman to the lead architect on the project. Often the cost of performing the paperwork on the project is as much as the personnel management for the field work.

In such cases both the contractor and the project engineer or project manager are obliged to follow the rules exactly. Of course there are still shortcuts to expedite the process of submittal and approval. These include standard forms for change orders, and approval methods. Submittal by electronic means or via email is one of the most promising methods for large government projects. This can take several forms in addition to regular email, which is not always very secure:

- Use of an Internet Web site to forward documentation. This is not always appropriate for document transfer as it is accessible to the public.

- Use of a bulletin board system (BBS). This is a precursor to the Internet, but it is appropriate for the secure transfer of documents over phone lines. The latest packages, notably a fine one from Wildcat!, is a ready-made network for interoffice communication. It has email, mail boxes, automatic downloading and uploading of documents, and even on-line conferencing. The best feature is that the system has a very sophisticated security (that Wildcat! claims has never been compromised) to limit access to very specific levels for all personnel logging on.

- Creation of an "intranet" Web site. This is one step beyond a BBS graphically, although the features are identical. Wildcat! has an Internet server upgrade to its workhorse BBS that creates a stand-alone Web site, complete with icons and all the other nice features we associate with Web sites. Essentially the user logs on via a Wildcat browser (which the licensed owner can distribute at no costs on a CD) and accesses the main Web site on the central computer. It can all be done on a good personal computer, over regular phone lines and using off-the-shelf technology elsewhere. (The intranet site can actually be made the user's actual Web site and can reside on the same computer by arranging the requisite connectivity with a local Internet provider.)

The idea of all these systems is to short circuit the paper trail. Instead of mailing, faxing, or sending documents by courier, it is all done instantly on-line. Here is how the system works during contract administration:

- The field contractor needs to contact the engineer urgently to clarify a question that is holding up progress. The detailed question is written up and emailed to the intranet server, to the engineer's attention. The server contacts the engineer's pager and notifies her or him of an important message, which is retrieved either via phone or on the computer.

- The contractor needs clarification of a part of the plans that was unclear. The engineer opens up the CAD drawing file, changes a thing or two, blows up the section in question, makes a separate 8.5- by 11-in drawing, and uploads the drawing with a written description attached, complete with a message to page the contractor to pick up the information.
- The contractor receives the page, goes to the field office, downloads the file, printing it out on the laser printer in the field office, and goes right to work.

The same kind of coordination can be done during the contract documents phase of a project, in coordinating work between architects and the engineering disciplines. Often several small consultants are doing different aspects of the project, in separate offices or even separate cities. Every time a new background, for example, is created by the architect, it must be delivered to the other consultants on the team. Instead of using disks, arranging a courier, and making check prints and a transmittal letter, the new background is simply uploaded to the intranet server and each consultant downloads it from there. The work of a day or two is done in 5 min.

In the case of change orders, the same method as just described applies. It is a quick and efficient means to resolve conflicts accurately and with authority. Each communication is kept on the server, too, so the entire record is on file for reference by anyone on the project staff from the owner on down (depending on their assigned access availability, of course). Other important project documents can also be retained on file for access as well as input by the entire team. Some things that can be done include

- Microsoft Project Management can be used to set up a project schedule to include all disciplines, field activities, reports, and deadlines. The schedule can be updated regularly with progress reports to track the whole project. The software also tracks worker hours, costs, and many other important project parameters. Having all of this information readily available will very much impress the client and will replace many meetings between all the parties. Allowing access and comment by other team members will help them to resolve conflicts in their own work by swiftly communicating their dilemma to all parties involved.
- Miscrosoft Scheduler can be used to keep a common calendar of project meetings and phone numbers for all personnel, contractors, and equipment suppliers.
- A current copy of all project documents, including drawings and specifications, is available at all times so that individuals can try to

answer their own questions before referring to someone higher up the chain of command. This documentation includes CAD drawings files, which can be opened and reviewed by many modestly priced software packages. Not only do the users not require electronic drafting skills, but their ability to do nothing more than read and print drawings keeps them from inadvertently modifying them.

When such a system is implemented on a large project, or between a general contractor and a group of subcontractors who often team together on projects, the lines of communication are always open. Change orders can be filled out, submitted, and returned with only a few keystrokes, no paper, and precious little time. Approval is swift and direct, and all parties involved can keep constant track of the scope and progress of the work.

One of the biggest advantages to the intranet concept is the availability of project drawings to all team members as well. Engineers and designers can access and work on the drawings during the construction document phase of the project, telecommuting. They will also have access to the drawings during the construction phase and can use them to make decisions after hours by accessing the intranet site from a remote terminal via a laptop computer on a regular phone line.

The best part about this whole system is that the hardware and software is inexpensive, easy to set up, and extremely dependable and rugged in operation. The user's role is even simpler. Any contractor who can navigate the Internet will be right at home.

Glossary

Access Fitting A mechanical fitting in an accessible location that permits easy access to conductors in a concealed space or to enclosed wiring, other than at an electrical outlet or light or other device mounted on a junction box.

Affinity Laws A set of formulas used to evaluate the operation of a pump or fan motor in a mechanical system. The most often quoted law is that power used is proportional to the ratio of motor speeds cubed.

Alternating Current (ac) An electrical current that alternates or reverses in direction at a specific frequency (typically 50 or 60 Hertz) and is configured as a sinusoidal waveform.

Alternator A synchronous alternating current device that changes mechanical power into electrical current.

Ammeter An instrument that measures, in amperes, the flow of electric current through a circuit.

Ampere or Amp (A) The measurement of electrical current that is equivalent to 1 V acting through a resistance of 1 Ω.

Amplification In a transistor circuit a small current controls a larger one, and the circuit configuration is used to increase the signal level.

Amplitude The height of the sinusoidal waveform for alternating voltage or current signal patterns.

American Wire Gage (AWG) A standard measure of wire sizes in terms of its diameter and cross-sectional area. Higher gage numbers indicate thinner wire sizes and the circular area doubles with every three gage sizes.

Analog Meter A measuring device configured so the pointer deflects in proportion to the current flow.

Arcing Contacts When an arc sparks between contacts, after the main contacts of a switch or circuit breaker have parted.

Arcing Time The time between the breaking of a fused link to the final interruption of current in the circuit, usually under specific conditions for a generic rating.

Armored Cable A cable wrapped with metal, typically steel wires, for the purpose of mechanical protection of the cable within.

Arrestor, Lightening A device that reduces the voltage of a lightening surge applied at its terminals and then restores the mechanism to the original conditions.

Attenuation Reducing the voltage in a current by a fixed proportion.

Audio Frequency Oscillator A circuit that is designed to generate sound in the audio range.

Auxiliary Contacts Extra contacts (other than the primary current connections) on a contactor used to introduce a signal to a control circuit.

Back emf A force that is created by alternating current in an inductor. It acts to oppose the force creating it. Also called the *counter emf*.

Bandwidth The frequency range of a resonating circuit, between the upper and lower half-power points.

Battery An electrochemical device that is used as a dependable source of direct current power.

Bleeder Resistor A resistor connected across the output of a filter circuit to discharge the capacitors when the power is cycled off.

Brake Horsepower The horsepower needed to operate a motor at a specific speed and power output.

Capacitance A property by which a system of conductors in a dielectric matrix is able to store electricity when a potential difference is applied between the poles.

Capacitive Filter A circuit that uses a capacitor connected in parallel across the load as a filter.

Capacitive Reactance The opposition created in a capacitive circuit that is energized by alternating current.

Cathode Ray Tube (CRT) A vacuum tube used in TV sets as a picture tube or as a monitor on computer terminals.

Center of Power A term used in applying the methods of Operations Research to the optimization of power distribution.

Center Tap A power connection to the center of a transformer.

Choke Another name for a coil of wire used as an inductor.

Circuit The path by which electrons travel through an electronic device.

Circular Mils (cmil) A measure of the cross-sectional area of round wire. One circular mil is the cross-sectional area of a wire with a diameter of 1 mil (.001 in). The number of circular mils in any size wire is the square of the diameter in mils.

Coaxial Cable A transmission line that transmits high frequency through a solid core, with a woven wire exterior conductor to act as a shield.

Conductor   A material that allows the movement of electrons, or electric current, with relatively little resistance. Examples of good conductors are copper, aluminum, and other metals.

Conduit   Metal or plastic pipe or tubing in which wiring is routed through a building, in the slab, or elsewhere on a job site. The conduit protects the wiring, organizes it, and makes it possible to replace wiring or pull new wiring without resorting to a major renovation.

Contactor   An on-off device for establishing or interrupting an electrical power circuit.

Contacts   Metallic conducting elements of a circuit that interact to complete or interrupt the connection.

Coefficient of Coupling   The effectiveness of the primary-to-secondary energy transfer in a transformer.

Component   A term used to describe small electronic parts.

Connector   A device that allows easy disconnection or connection of a device from a circuit.

Cycle   A term used to describe frequency (e.g., hertz is cycles per second).

Deflection Plates   Elements shaped like plates that are used to control the beam position as it moves from side to side in a cathode ray tube.

Dielectric   An insulating material that can be subjected to electrical stress, a property that allows it to store charge.

Diode   A two-terminal electronic device that allows current to flow in only one direction.

DIP (Dual in-line package)   A term used to indicate the packaging of integrated circuits.

Direct Current (dc)   A constant value electrical current that, unlike alternating current, flows only in one direction.

Direct Digital Control (DDC)   A term descriptive of modern controls systems that are based on electrical devices, electronic control strategies, and transistorized circuitry.

DMM   Abbreviation for digital multimeter.

Dry Cell   A cell consisting of zinc and carbon with an electrolyte of salt-ammonia compounds.

DX (direct expansion)   Used to describe hvac units that use a refrigerant and a small dedicated compressor instead of refrigerant in a central chiller.

Dynamic Head   Pressure losses in a hydronic system due to flow.

Eddy Current   The current induced in the core of a transformer, inductor, or solenoid that is caused by the variable magnetic field of alternating current.

Effective Voltage   Another term for root-mean-square (rms) voltage, or .707 times the peak voltage of an alternating current circuit.

Electrical Interlock A wiring method that, when one device is energized, a control circuit is opened in one or more other devices, causing them to open or preventing them from closing.

Electromagnet A magnet that is created by the flow of electric current through a coil of wire, often with an iron core.

Electromagnetism The creation of a temporary magnetic field by the flow of electrons through a wire.

EMS (Energy Management System) An all encompassing term that describes the overall controls harness for a building, for hvac, electrical, lighting, security, and fire alarm systems.

Farad (F) The unit of measurement for capacitance.

Feeder A major electrical conduit, usually between a distribution panel and a remote electrical panel. The term is also used for conduits to large hvac or other equipment of high voltage and/or current-carrying capacity.

Filter A special circuit connected across a power supply to reduce the distortions caused by the full- or half-wave rectification of the alternating current.

FLA (Full Load Amps) The current drawn by a motor or other device at rated speed or normal, steady-state conditions.

Flux The magnetic field of a coil.

Flux density The number of magnetic field lines per unit area. The greater the number of lines, the stronger the magnetic field induced.

Frequency The cycles per second (or hertz) of an alternating current circuit.

Friction Head The dynamic head increases in proportion to the square of the flow rate in most elements of a fluid system, such as pipes, valves, and fittings.

Full-Wave Bridge Rectifier A transistor device that uses four diodes in a diamond-shaped bridge arrangement to cause full-wave rectification of alternating current. The output is a steadier current than is created by a two-diode circuit.

Full-Wave Rectifier A device that uses two diodes to create alternating current.

Fuse A circuit protection element that melts to break the circuit when the current exceeds the rated value.

Gain The proportional increase in voltage or current produced in an electronic circuit, especially transistor devices.

Galvanic Corrosion When two dissimilar metals are in contact, an electric potential can develop between them and in time result in a physical corrosion of the connection.

Generator An electromechanical device, often with an independent or emergency power source, used to produce electrical power for commercial use.

Ground A connection to the earth, or a common connection point in an electronic device with a known potential.

Grounding To establish a conducting connection between a circuit or equipment and the earth.

Half-Wave Rectification A single diode converts half of the alternating current cycle into direct current that is pulsating. Combinations of two or four diodes result in a more complete conversion of the ac power to dc, with a more uniform wave pattern.

Harmonic An even-numbered multiple, usually in terms of frequency, of a given waveform.

Head A measurement of fluid pressure in a hydronic system, usually in feet of water. A 30 ft head, for example, is equivalent to the pressure at the bottom of a column of water 30 ft high. Commonly divided into static and dynamic head.

Heat Sink A device used to dissipate heat from electronic or other types of components. Large metal fins on diodes are one example. They dissipate heat generated during operation of the device.

Henry The unit of measurement for inductance.

Hertz A measurement of frequency, equal to one cycle per second.

Home Run The conduit run from the point(s) of use to the panel. The term is usually applied to a single circuit, even though the actual conduit often carries the wiring for several independent circuits.

Horsepower (Hp) A measurement of power, or the capacity of a device to perform useful work. 1 Hp = 746 W.

hvac (heating, ventilating, and air conditioning) An abbreviation for the standards, equipment, and installation of air conditioning systems for the building industry.

Hysteresis The magnetic energy lost or dissipated in a transformer or inductor that is the result of the residual magnetism that remains in the material after each cycle of the imposed electromagnetic waveform.

Impedance Symbolized by the letter Z, it is the total opposition to the flow of current in a circuit as caused by the combined effects of resistance, capacitance, and inductance.

Impedance Matching The selective use of equipment to cause the most complete and efficient transfer of energy. An example is transformers matched to amplifiers.

Impedance Ratio The relationship between the primary and secondary windings of a transformer.

Induced Current Whenever magnetic field lines cut across a conductor, a current is induced in the conductor.

Inductance A circuit element that tends to oppose any change of current because of the magnetic field associated with the current carrying conductor.

Insulator A material that prevents the movement of electrons, or electric current, such as plastic, rubber, or a similar synthetic material.

Integrated Circuit An electronic device that is the collection of many transistors combined in a single circuit element and manufactured as one product on a chip.

Kirchoff's Current Law The algebraic sum of currents at a junction equals zero, so the current flowing into a junction equals that leaving the junction.

Kirchoff's Voltage Law The algebraic sum of voltage drops around a closed circuit equals zero, or the voltage drops around a loop equals the applied voltage.

Law of Magnets Like poles repel and unlike poles attract.

Life Cycle Cost A method of comparing to real cost of several different alternatives that includes initial, operating, maintenance, and replacement costs. A thorough evaluation includes the time value of money, escalation in utility costs, and other pertinent factors—especially if the investment is a large one.

Light Emitting Diode (LED) A semiconductor device that glows when it is forward biased. LEDs are used to form digital displays on many devices.

LRA (Locked Rotor Amps) The amount of in-rush current drawn by a motor when it is first started, before it begins steady-state operation and draws full load amps (FLA).

Magnetic Field A force that surrounds a magnet and that is silhouetted by arranging metal filings into patterns that are the lines of magnetic flux.

MDP (Main Distribution Panel) The largest electrical panel in a facility, usually near the service entrance, from which remote panels, equipment branches, and all other major equipment is fed.

Mechanical Interlock A device that keeps two contactors from being closed at the same time, thus causing a short in the line.

Modulation The process by which an audio or other frequency is impressed upon a carrier wave for transmission over long distances.

Mutual Inductance Inductance that occurs when adjacent conductors such as windings of a transformer interact.

NEC (National Electric Code) The standard for electrical installations in the United States.

Ohm (Ω) The measurement of electrical resistance, or a potential drop of 1 V when the current is 1 A.

Ohm's Law The relationship by which current in a circuit is inversely proportional to the resistance and voltage and current are directly proportional in a circuit.

Open Circuit When the flow of current to a circuit is interrupted, either unintentionally or as the result of a switch or contactor.

Open Transition When a motor is started at reduced voltage, a transition must be made when it reaches full speed and load to full voltage input. When this occurs, there are 20 or 30 ms when power is disconnected from the motor completely. As it is reconnected at full voltage, there is a spike in the line current.

Operations Research The analytical practice of determining the shortest, most efficient path between multiple points.

Operating Point The intersection of the pump curve and the system curve.

Phase The difference between 0-V crossings of two signals, expressed usually in degrees and sometimes termed a phase shift.

Power Voltage times current, measured in watts.

Power Factor The ratio of active power in kilowatts to apparent power kilovoltamperes in a circuit.

Present Worth An economical analysis term useful in evaluation or cost estimating of equipment. The method estimates the annual costs, using the time value of money and a projected rate of inflation, and calculates the value of the expenditures in current funds.

Pump Curve The characteristic curve of a pump showing the pressure head versus flow relationship.

Replacement Cost The total cost to replace an item or system, including initial cost and annual maintenance and operating costs.

Residual Magnetism What remains after the magnetizing force has been removed.

Resistance The opposition of a material to the flow of electrons through it.

Resistivity The rate at which energy is dissipated from an electron flow through a wire. This factor is independent of the length or the area of the wire, so it compares different materials on equal terms.

Resonance When the capacitive reactance is the same as the inductive reactance in a circuit.

Rheostat A two-terminal device that serves as a variable voltage control point.

Root-Mean-Squared (rms) An average of the voltage or current in alternating current systems that is 0.707 times the peak voltage or current.

Rotor The moving part of an electric motor or generator.

Salvage Value The remaining useful value, in terms of money, that exists when a piece of equipment or a system is removed from service and replaced. The salvage value of the existing system is added to the cost of the new system, when the relative merits of the new and old systems are compared.

Series Circuit A circuit arrangement in which all the current flows through every part of the circuit.

Short Circuit When two active electrical conductors are cross connected, resulting in a very strong electric current through the conductors.

Shunt A synonym for parallel; for example, a resistor placed in parallel with an ammeter meter movement to extend its range of sensitivity.

Specific Resistance A relative measurement of resistivity in wire based on a standard wire length of 1 ft and a cross sectional area of 1 cmil.

Static Electricity Electricity generated by friction. It remains in place until a path is provided for it.

Static Head The pressure needed to overcome a change in elevation in a hydronic system, expressed in feet of water.

Stator The fixed element in a motor or variable capacitor.

System Head Plot of the pressure needed to balance the static and the dynamic head for a range of flows in a hydronic system.

Terminal Strip A series of connection points that are insulated from each other and that provide places for components to be wired or joined.

Tolerance The amount of error that can be permitted between the actual and indicated values of a component.

Timing Relay A circuit element that causes a fixed, preset delay in the control sequence.

Turns Ratio The relationship between the turns on the primary and secondary sides of a transformer.

VAV (variable air volume) An hvac system that can deliver differing volumes of air to each conditioned space, in proportion to the actual space temperature with respect to the target or setpoint temperature.

Volt (V) A measure of electromotive force. One volt of potential difference is created when 1 A of current flows through a circuit with a resistance of 1 Ω.

Voltampere (VA) A measure of the apparent power of a device, or its true power capability.

Voltage Divider Resistances arranged in series to create a proportionately lower amount of voltage by tapping across different resistor combinations.

Voltmeter An instrument that measures the potential difference, or volts, between two points in a circuit.

Watt (W) A measure of electrical power that is produced in an electrical circuit when a potential difference of 1 V causes a current on 1 A to flow.

WSHP (water source heat pumps) A packaged air conditioning system that discharges heat to a circulating fluid rather than to air at an outside condensing unit.

GENERAL NOTES:

1. ALL ELECTRIC OUTLETS IN KITCHEN TO BE GFI
2. ~~A COMPLETELY NEW LIGHTING/POWER SYSTEM~~ TO BE PROVIDED. EXISTING SWITCHES AND OUTLETS CAN BE THEY MUST BE REPLACED WITH NEW ITEMS ACCEPTABLE TO CURRENT CODE REQUIREMENTS.
3. PROVIDE NEW 200A SERVICE, COMPLETE WITH WEATHER HEAD, GROUNDING ROD, CT-CAN AND SERVICE TO THE MAIN HOUSE PANELL. PROVIDE THE FOLLOWING BREAKERS IN THE HOUSE PANEL FOR SPECIAL APPLIANCES.

   40A/2O CB FOR CONDENSING UNIT
   20A/2P CB FOR AIR HANDLER
   10 20A/1P CB'S FOR POWER
   5 20A/1P CB'S FOR LIGHTING
   5 20A/1P SPARES

4. INSTALL NO MORE THAN 7 DUPLEX OUTLETS PER CIRCUIT; NO MORE THAN THREE PER CIRCUIT IN THE KITCHEN.

KEYED NOTES:

1. NEW SERVICE, WEATHER HEAD
2. NEW HOSE BIB
3. NEW ELECTRIC PANEL
4. TRACK LIGHTS
5. CONDENSING UNIT DISCONNECT
6. DISHWASHER
7. GARBAGE DISPOSAL (J-BOX)
8. REFRIGERATOR (DEDICATED DUPLEX OUTLET)
9. GAS STOVE, WITH J-BOX FOR CLOCK, IGNITION
10. GAS WATER HEATER, WITH ELECTRONIC PILOT LIGHT
11. GAS DRYER (NO POWER)

WASHING MACHINE (220V OUTLET)

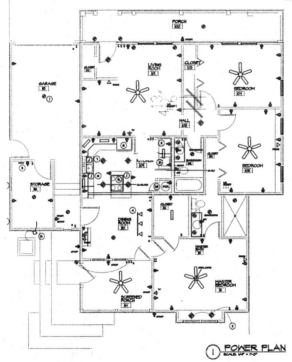

Figure G.1 Residence—power and lighting.

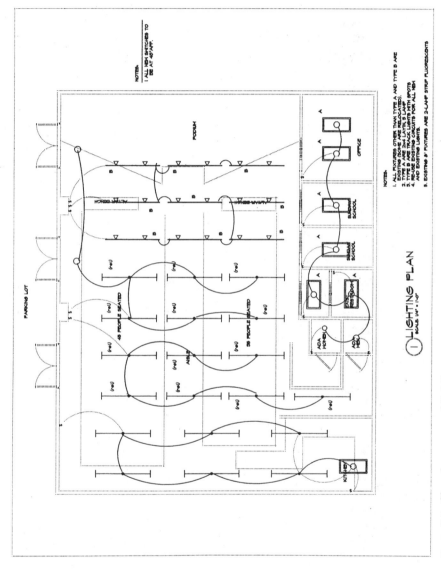

Figure G.2 Chapel—power and lighting.

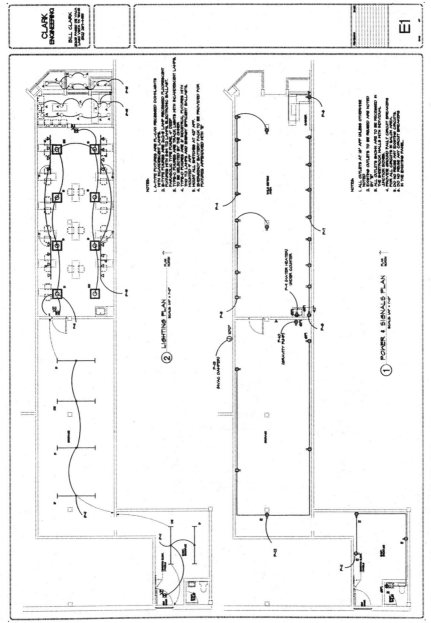

Figure G.3 Retail—power and lighting.

Glossary

GENERAL NOTES

1. UNLESS OTHERWISE SHOWN, ALL BRANCH CIRCUIT WIRING IS 3/4" CONDUIT CONTAINING 2#12 CONDUCTORS (PLUS #12 GROUNDING CONDUCTOR THAT SHALL BE IN ALL BRANCH CIRCUITS REGARDLESS OF INDICATION). IF NOT INITIALLY INSTALLED, CONTRACTOR SHALL BE REQUIRED TO INSTALL GROUND CONDUCTORS AT HIS OWN EXPENSE.
2. REFER TO ARCHITECTURAL & MECHANICAL PLANS FOR EXACT LOCATION OF EQUIPMENT.
3. WHERE ELEVATIONS OF ELECTRICAL OUTLETS ARE SHOWN ON DRAWINGS, THEY ARE GIVEN AS AN AID TO THE CONTRACTOR FOR BIDDING AND ROUGH-IN. COORDINATE FINAL EXACT LOCATION OF ALL OUTLETS WITH ARCHITECTURAL PLANS, ELEVATIONS & CONSTRUCTION DETAILS.
4. THE WORD "(ON)" NOTED ADJACENT TO A HOME RUN CIRCUIT NUMBER INDICATES THAT THERE ARE ADDITIONAL HOME RUNS OF THE SAME CIRCUIT FROM OTHER AREAS. GROUP & CONNECT COMMON HOME RUN WIRING PRIOR TO ENTERING PANELBOARD ENCLOSURE.
5. NO THRU THE WALL BOXES SHALL BE INSTALLED WHERE OUTLETS OCCUR BACK-TO-BACK IN A COMMON PARTITION.
6. ALL EMPTY CONDUIT SHALL CONTAIN A PULL CORD. ALL EMPTY CONDUIT STUBBED INTO ACCESSIBLE CEILINGS SHALL BE TERMINATED ABOVE CEILING WITH A PLASTIC BUSHING.
7. PROVIDE 3/4" CONDUIT WITH PULL CORD FOR EACH TELE/COMM OUTLET. ROUTE TO NEAREST ACCESSIBLE ATTIC OR CEILING SPACE WITH A PLASTIC BUSHING.
8. SECURITY ACCESS SYSTEM: (1) CARD READERS OF proximity TECHNOLOGY, SINGLE STAGE, COMMUNICATING WITH THE FIELD PANEL VIA 4-CONDUCTOR 22AWG WIRE, UP TO 3000 FT. AWAY, WITH VISUAL AND AUDIO SIGNAL ALARMS, (2) U.L. LISTED RIM DEVICE, 4' WITH SIGNAL SWITCH, (3) MAGLOCKS, RUTHERFORD MODEL 8510 FOR SINGLE DOORS, 8520 FOR DOUBLE DOORS, WITH DOOR STATUS SENSOR AND LOCK STATUS SENSOR; (4) INTELLIGENT PANEL, UL 244/1076 LISTED WITH MODEM AND PC INTERFACE. ALL ITEMS TO BE PROVIDED BY THE SAME MANUFACTURER. PROVIDE 20A CB IN NEAREST PANEL FOR INTELLIGENT PANEL.

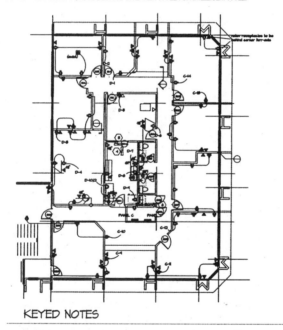

KEYED NOTES

(1) EXISTING 225 A PANEL. INSTALL J-BOX AT LOCATION ABOVE CEILING, THEN ROUTE NEW FEEDER TO PANEL A, 4-#4/0, #26 IN 3"C.

(2) EXISTING 400 A PANEL - REMOVE. INSTALL J-BOX AT LOCATION ABOVE CEILING, THEN ROUTE NEW FEEDER TO PANEL B, 2 SETS OF (4-#4/0, #26 IN 3"C.)

(3) EXISTING 225 A PANEL. INSTALL J-BOX AT LOCATION ABOVE CEILING, THEN ROUTE NEW FEEDER TO PANEL C, 4-#4/0, #26 IN 3"C.

(4) EXISTING 400 A PANEL - REMOVE. INSTALL J-BOX AT LOCATION ABOVE CEILING, THEN ROUTE NEW FEEDER TO PANEL D, 2 SETS OF (4-#4/0, #26 IN 3"C.)

(5) EXISTING TELEPHONE BOARD. REMOVE

(6) PROVIDE 4'x 8' PLYWOOD BOARD WITH BUILDING GROUND. PAINT WITH TWO COATS OF NON-CONDUCTING PAINT.

Figure G.4 Office building—power.

Index

Accelerating torque, 158
ADA (Americans with Disabilities Act) devices, 71
Addenda, 333
Air-cooled chillers, 136
Air handler rooms, 126
Alternating current, 30
Aluminum cable, 6
American Wire Gage (AWG), 4
Angular frequency, 35
Apparent power, 38
Area factors, 313
AutoCad, 270

Balancing loads, 62
Bidding, 332
Boilers, 131
Bridge sensitivity, 26
Bulletin board software, 336

Capacitors, 33
Central plant, 128
Change orders, 335
Checklist, 299–302
Circular mils, 4
Communication rooms, 68
Complex plane, 32
Computer applications, 195
Computer networking, 72
Conducting metals, 3
Conduit costs and labor values, 189
Constructability, 210
Construction management, 334
Control valves, 139
Controls, 135

Cooling tower, 129
Core losses, 170
Corrosion, galvanic, 6
Costs, 213
Cove lighting, 82
Cover requirements, minimum, 13
Current limiters, 320

Database programs, 196
Daylighting strategies, 108
Dead-end attachment, 49
Delta-connected load, 61
Delta connections, 175
Demand factors, 99
Derating, 147
Design development, 328
Desuperheaters, 136
Device labor values, 181
Dimming controls, 86
Disconnect labor values, 188
Disconnect switch, main, 52
Duty cycling, 236

Eddy currents, 171
Electric heating, 240
Electromagnetic theory, 88
Energy management system, 266
Equipment grounding conductors, 140
Estimating, 179
Exhaust fans, 132
Exit signs, 65, 323

Feeder size, 12–13
Fiber optic cabling, 72

352 Index

Fire alarm devices, 69
Fire alarm riser, 70
Fixture budget, 106
Fluid analogies, 2,8
Frequency domain, 39
Full-load current data, 150

General notes, 277
Generator sizing, 143
Grounding conductors, 139
Grounding electrode conductor, 140

Hanger lengths, 116
Harmonics, 39, 173
Heat pumps, 135
Horsepower, 154
Hot water storage, 267
Humidity control, 263
Hysteresis losses, 172

Illuminance categories, 81
Impedance, 35
Indoor air quality, 237
Inductors, 33–24
Inertia, 159
Insulators, 11, 14–16
Intranet Web site software, 336
Isolation transformer, 178
I-squared R losses, 7, 10, 20

Junction box specifications, 51

K-factor transformers, 245
Kirchoff's laws, 27–29
Kitchen layout, 121

Lamp lumen depreciation, 113
Lamp selection, 105
Light circuit design, 85
Light circuiting, 84
Light levels, 79–80
Light shelf, 109
Lighting and circuiting, 77
Lighting energy conservation, 87
Lighting feeder demand factors, 101
Lighting plan, 327
Lighting power density limits, 314–319

Lighting power limits, 311
Load analysis form, 102
Load control, 235
Load estimating, 46
Luminaire selections, 78
Luminous efficacy, 117

Maximum locked rotor current, 148
Mechanical room layout, 123
Mechanical throttling, 163
Meter loop, 44
Motion sensors, 86
Motor circuiting, 167
Motor coordination, 151
Motor demand factors, 100
Motor gensets, 246
Motor labor values, 182
Motor laws, 153
Motor starting KVA, 144
Multiple meter installation, 47

Neutral current, 29
Nominal motor efficiencies, 319
Nonlinear loads, 242
Nonmetallic cable, 12

Ohm's law, 19
One-line riser diagrams, 55, 57, 58, 60
Operations research, 91
Optical reflectors, 321
Outside air systems, 255
Overhead service clearances, 49
Overload protection, 142

Packaged rooftop equipment, 124
Panel labor values, 192
Panelboard schedule, 60
Parallel circuit, 22
Peak building loads, 147
Peak demand reduction, 268
Phasor representation, 61
Points of use, 52
Polarity, 176
Polyvinyl chloride (PVC), 11
Power and signals plan, 326
Power consumption by appliances, 52
Power factor correction, 40–41
Power quality, 241
Power receptacle densities, 54

Proactive controls, 237
Production process, 323
Project management software, 337
Project scheduling software, 337
Pump curves, 162
Pumps, 132
PVC (polyvinyl chloride), 11

Quality control, 330

Reactance, 36
Reactive power, 37
Rebate programs, 312
Reduced voltage starting, 145
Reflectance factors, 114
Reflected ceiling plan, 80
Resistance losses, 8
Resistance, 3–5
Roof ventilators, 131
Running overload units, 142

Safety factor, 1
Schedules, 57, 64, 90, 168, 178
Schematic phase, 324
Security devices, 74
Series circuit, 21
Series-parallel circuit, 24
Service conduit, 50
Service corridors, 180
Service drop conductors, 48
Service entrance cable, 12
Service entrance conductors, 47
Service entrance, 48
Service installation, 43
Service size table, 45
Short circuit, 23
Shunt balance, 26
Sinusoidal waveforms, 31
Site plan, 325
Spacing and mounting heights, 115
Specifications, 281–297, 304–310
Split systems, 125
Starter labor values, 184

Steel raceway uses, 56
Submetering, 265
Symbols, 59, 63, 76, 89
Synchronous motors, 248

Telephone cables, 66
Telephone/data symbols, 67
Temporary service, 43
Thermal storage, 252
Three-phase ac motor data, 149
Three-phase power, 29
Three-phase zoning, 95
Time clocks, 138
Torque, 156
Transformer efficiency, 174
Transformer labor values, 186
Transformer, 169
Trench detail, 51
Two-speed motors, 160

Uninterruptible power supply (UPS), 244
Utility yard layout, 122

Value engineering, 209
Variable air volume (VAV), 125
Variable-frequency drives (VFDs), 127, 134
Variable torque loads, 165
VAV (variable air volume) boxes, 130
Visual environment, 103
Voltage drop, 7, 17
Voltage drop program, 201
Voltage ratio, 169

Water heaters, 133
Water source heat pumps, 137, 252
Whetstone bridge, 25
Wire pulling labor values, 190
Wye connection, 177

Zonal cavity calculations, 111

ABOUT THE AUTHOR

William H. Clark II is a registered professional engineer with wide experience in electrical, lighting, mechanical, and structural projects. An expert on energy-efficient materials and designs, he is the author of *Retrofitting for Energy Conservation*. He is a member of the Technical Calculations Committee for the Illuminating Engineering Society, and a member of IEEE, ASHRAE, and ASME, as well as a member of the board of directors of *Engineered Systems* magazine. Mr. Clark has also written articles for numerous trade and technical journals, as well as several computer programs to model energy conservation strategies.